Edward Wheeler Scripture

Thinking, Feeling, Doing

Edward Wheeler Scripture

Thinking, Feeling, Doing

ISBN/EAN: 9783743402744

Manufactured in Europe, USA, Canada, Australia, Japa

Cover: Foto ©berggeist007 / pixelio.de

Manufactured and distributed by brebook publishing software (www.brebook.com)

Edward Wheeler Scripture

Thinking, Feeling, Doing

THE AMERICAN FLAG
AS SEEN
I.-BY MOST PEOPLE.
II.-BY RED-BLIND PERSONS.
III.-BY GREEN-BLIND PERSONS.
IV.-BY VIOLET-BLIND PERSONS.
V.-BY TOTALLY COLOR-BLIND PERSONS.

THINKING, FEELING, DOING

BY

E. W. SCRIPTURE, PH.D. (LEIPZIG)

Director of the Psychological Laboratory in Yale University.

FLOOD AND VINCENT
The Chautauqua-Century Press
MEADVILLE PENNA
1895

THE AMERICAN FLAG
AS SEEN

I.- BY MOST PEOPLE.
II.- BY RED-BLIND PERSONS.
III.- BY GREEN-BLIND PERSONS.
IV.- BY VIOLET-BLIND PERSONS.
V.- BY TOTALLY COLOR-BLIND PERSONS.

THINKING, FEELING, DOING

BY

E. W. SCRIPTURE, PH.D. (LEIPZIG)

Director of the Psychological Laboratory in Yale University

FLOOD AND VINCENT
The Chautauqua-Century Press
MEADVILLE PENNA
1895

Gift of
Miss Agnes Stowell

The Chautauqua-Century Press, Meadville, Pa., U. S. A.
Electrotyped, Printed, and Bound by Flood & Vincent.

PREFACE.

A FELLOW PSYCHOLOGIST said to me one day, "Are you not afraid that all this accurate and fine work in the laboratory will scare away the public?" This book is the answer. You, my dear reader, and I, have no time, inclination, or means to spend years in studying the details of the physical laboratory or the observatory, yet we both enjoy an account of the latest advances of electricity by a specialist in physics or a series of new photographs of the moon by an astronomer. Life is so short that a man can learn only one thing well, whether it be the best method of dyeing cambric or the most efficient construction of locomotives. The botanist is quite at home with the plants but is ordinarily as ignorant of psychology as a stock-broker—if not more so. The mathematician learns some section of mathematics, but would be just as awkward at a chemical analysis as any other outsider. We all belong to the great public except in regard to the particular handiwork, trade, or science that each knows something about. And yet we are all interested in hearing about a new science. There is nothing too good for the public—for you and for me; the finer the work, the more novel the invention, or the more important the discovery, the greater the duty of telling it to the public in language that can be understood.

The greatest of psychologists, Wundt, has written a series of lectures on psychology (lately translated into English), but the style and the matter are fully intelligible only to those who are already somewhat familiar with the science. No one else has produced a book explaining the methods and results of the new psychology. This is my reason for writing one.

This is the first book on the *new*, or experimental, psychology written in the English language. That it has been written *expressly for the people* will, I hope, be taken as evidence of the attitude of the science in its desire to serve humanity.

In one respect I have departed widely from the usual writers on psychology; I have written plain, every-day English and have not tried to clothe my ignorance in the "multitudinous syllabifications and frangomaxillary combinations" that pass as philosophic English.

CONTENTS.

LIST OF ILLUSTRATIONS.

THINKING, FEELING, DOING.

THINKING, FEELING, DOING.

CHAPTER I.

WATCHING AND TESTING; OR OBSERVATION AND EXPERIMENT.

EYES and No-Eyes journeyed together. No-Eyes saw only what thrust itself upon him; Eyes was on the watch for everything. Eyes used the *fundamental method of all knowledge—observation, or watching.* **The method of acquiring knowledge.**

This is the first thing to be learned—the art of watching. Most of us went to school before this art was cultivated, and, alas! most of the children still go to schools of the same kind. There are proper ways of learning to watch, but the usual object lessons in school result in just the opposite. We, however, cannot go a step further till we have learned how to watch.

Do you wish to know just how your children play together? Watch them, but watch them so that they do not *feel* your presence. **The fundamental law of watching.**

Every public man wears a mask, because he is watched. If we wish to know just what *kind* of a man he is, we must watch him in unsuspected moments. A great deal of ridicule has been cast on the reporters and enthusiasts who rush into the room that has just been occupied by a great man in order to see how he has left the chairs, how he has treated the curtains, how much soap he has used, or how many towels he has soiled; or who interview the cashier to find out just how many cocktails he has consumed. In one respect these men are quite in the right. They say to themselves, "The public is interested in knowing just what the man really is when he has his mask off, and that is only when he is alone." The man who thoughtlessly leaves behind a soiled deck of cards, a whisky bottle, and the odor of bad cigars must be quite a different fellow from one who has had an artistic dinner and a copy of the latest novel, or one who has left his Bible and his spectacles on the table.

If there is anything wrong about this, it is not the *method;* just this method is to be used in acquiring all knowledge. In fact, I shall want you to watch the processes of thinking, feeling, and doing, in exactly the same fashion. Lie in wait, concealed, catch your "process" going on in a perfectly natural way. Moreover, strange as it may seem, this is the *only* way, the fundamental rule being that *the act of watching must not change the person or thing watched.*

The error of prejudice.

It is not sufficient to know this rule; we must be constantly on guard against several very dangerous sources of error. The first is the error of prejudice. Grandmother M. has used Dr. Swindle's liver pills all her life long. She always believed they would do her good; she remembers the dozen times she happened to feel better after taking them and forgets the hundreds of times she did not. Therefore she has facts—incontestable facts—to prove the goodness of the pills. Possibly her picture appears in the newspaper with an enthusiastic testimonial. It is useless to attempt to convince her that her method of observation has been vitiated by the *error of prejudice.*

Its presence everywhere.

Of course, this error is very plain in other people, but you, my dear reader, always judge fairly. Let me whisper in your ear: Have you not some pet fad on which you are sure you are right and all the rest of the town are wrong? are you not quite sure that there is only one side to the tariff question? are you not astounded at the fact that some people find a good side to a man you know—yes, *know*—to be utterly bad? Don't be ashamed to confess. The great scientist Faraday did. "It is my firm opinion that no man can examine himself in the most common things having any reference to him personally or to any person, thought, or matter related to him without soon being made aware of the temptation to disbelieve contrary facts and the difficulty of opposing it. I could give you many illustrations personal to myself about atmospheric magnetism, lines of force, attraction, repulsion, etc." If Faraday could go wrong in this way, how careful must we be in the observations we shall make, in the experiments we are about to perform.

Faraday's confession.

Another very dangerous *error* is that of *unconscious additions.*

Unconscious additions.

Play the game of twenty questions. The company choose some object and some one who does not know what has been chosen has to guess it from the answers "Yes" or "No" to his questions. Stop him when he is half through and ask him to tell you what he has concluded from the different answers. You will find that he adds far more than is justified by the answer to each question. For example, something chosen is neither animal nor mineral; it is, therefore, so the questioner thinks, "a" vegetable. But suppose you had chosen "buckwheat cakes"?

This error is one of the most troublesome ones in reading printer's proof; letters and words that have been omitted by the compositor are unconsciously supplied bv the reader. An author, on account of his interest, is more liable to this error than any one else ; he is generally a very unreliable proof-reader.

A familiar case of this error is found in the story of the ten white crows—which I will leave the reader to hunt up in his old school books.

Examples of unconscious additions in animal psychology.

This source of error, as Wundt has pointed out, renders almost absolutely worthless an enormous amount of painstaking work in animal psychology. The facts are observed and collected with untiring diligence, but the critical study of the results is generally entirely lacking.

Take, for example, a case reported by Romanes in his volume on animal intelligence.

"Funereal habits" of ants.

An English clergyman writes concerning the "funereal habits" of ants: "I have noticed in one of my formicaria a subterranean cemetery, where I have seen some ants burying their dead by placing earth above them. One ant was evidently much affected, and tried to exhume the bodies; but the united exertions of the yellow sextons were more than sufficient to neutralize the effort of the disconsolate mourner."

Wundt asks, How much is fact, and how much imagination? It is a fact that the ants carry out of the nest, deposit near by, and cover up dead bodies, just as they do anything else that is in their way. They can then pass to and fro over them without hindrance. In the observed case they were evidently interrupted in this occupation by another ant, and resisted its interference. The cemetery, the sextons, the feelings of the disconsolate mourner, which impelled her to exhume the body of the departed—all this is a fiction of the sympathetic imagination of the observer.

Another friend of ants gives this account : "At one

formicary half a dozen or more young queens were out at the same time. They would climb up a large pebble near the gate, face the wind, and assume a rampant posture. Several having ascended the stone at one time, there ensued a little playful passage-at-arms as to position. They nipped each other gently with the mandibles, and chased one another from favorite spots. They, however, never nipped the workers. These latter evidently kept a watch upon the sportive princesses, occasionally saluted them with their antennæ in the usual way, or touched them at the abdomen, but apparently allowed them full liberty of action."

The case of the young ant-princesses.

The correctness of this observation, says Wundt, need not be questioned. Why should not a number of young queens have been crowded together upon a pebble, and some workers have been with them, and occasionally touched them with their antennæ, as ants do everywhere? But that they "sported" and played, that the others "kept watch upon them" like chaperones, and now and again did homage to them by "saluting"—all this is due to the imagination of the observer. He would hardly have told the story in this way had not the suggestive name "queen" been introduced for the mature female insects. If the adults are "queens," the young ones must, of course, be "princesses" to the other ants as well as in the imagination of the observer. And since no princess ever went out without an attendant or a chaperone, the rest of the tale follows as a matter of course. If, instead of the name "queen," the mature female ant had been called by the still better term "mother," we would have had an entirely different story from the same facts. I leave it to my readers to tell it.

Wundt's remarks.

Watch Fido, the pet dog, at play. Let your friend

An experiment for the reader.

tell the story of what he was doing ; then tell it yourself. Notice how you both add your own imaginations to the facts. The story as told by a sharp business man, accustomed to beware of imagination, will be quite different from that of a lady novel-reader steeped in romance.

Need of experiment.

How easy it is to misinterpret an observation if the very greatest care is not taken in recording it, and if it is impossible to vary the circumstances by experiment and thus to obtain accurate knowledge of the details, is well shown by the following facts.

Pierre Huber, one of the most reliable students of the habits of ants, stated that he had assured himself that an ant, if taken from the nest and returned after an interval of four months, was recognized by its former companions ; for they received it in a friendly manner, while members of a different nest, even though they belonged to the same species, were driven away. The correctness of the observation cannot be doubted ; it has also been confirmed by Lubbock. Lubbock, however, made the matter a subject of experiment. He took ant larvæ from the nest and did not put them back till they were fully developed. They, too, were received in a friendly manner, although there could be no question of resemblance between the larva and the grown ant. There must, therefore, be some characteristic peculiar to all members of a particular nest, possibly a specific odor, which determines the "friendliness" of the ants.

Personal application.

Every one of my readers is an observer in a particular domain of mental life ; and, I fear, commits this error daily.

A pack of noisy boys is at play on the street. Ask a crusty old bachelor to tell what they are doing. Then ask "mother," who has had boys of her own, to tell the story. You will be surprised to learn from the former what villains those boys show themselves to be by their acts, whereas "mother" will point out to you how every movement of the boys proves them manly fellows.

By the way, I happen to notice that the expression by which I have introduced the boys to my readers contains such an error of prejudice that they can readily guess the sort of description I would write. Suppose I said and felt, "A group of merry boys"; would not my account of the very same facts be different?

Another source of error.

I shall warn you against only one *error* more, that of *untrustworthiness of the senses*, as it is called. Sir Walter Raleigh was one day sitting at a window when he observed a man come into the courtyard and go up to another standing by the door. After a few words the latter drew his sword, they fell to fighting, and the first comer was finally wounded and carried out. A person who had been standing close beside the door afterwards flatly contradicted the observation of Sir Walter, saying that the man at the door had not been the first to draw his sword and that it was not the first comer who was wounded and carried out. Note the flat contradictions of eye witnesses in the next trial you read about.

Lessons in observing.

Let us now take a few lessons in observing.

Exercise I.

1. On page 22 of this book—do not turn to it till I have told you what to do—you will find a figure. Write what you see. I am not going to tell you another thing about it; not even what the exercise is for. Show the figure to other people with the same directions. Compare your result with theirs. Just as you progress in understanding what the exercise is for, just so far will you have profited by it.

2. On the second page from this you will find a number of letters printed in a square. Turn over the page for just an instant and then close the book. What letters can you remember? You can readily prepare a set of cards with various combinations of letters and can train

Exercise II.

your friends in observing. Or you can use cut letters, such as go under the name of letter-tablets. Make irregular combinations on the table behind a screen of

Fig. 1. An Exercise in Observation.

some kind, *e. g.*, a book; snatch the book away for an instant, and have the onlookers write down the ones they saw. Then form words instead of letters. You will notice that people can catch almost as many words as they can catch disconnected letters. Or you can write on a slate and turn it over for an instant. Or you can use dominoes. The Italian game of "morra" is for this very purpose. One person holds up a number of fingers suddenly for an instant; the other guesses how many were shown.

Exercise III.

3. Place a number of objects on a table in the next room. Let each person go in and walk once around the table during the time you count twenty. Coming out he is to write down a list of what he saw.

The magician's training.

At first you can catch almost nothing in these last two exercises. It is very important to continue the practice; you cannot go too far. You will be encouraged by knowing that the magician Robert-Houdin began in the same way. He and his son would pass rapidly by a shop-window and cast an attentive glance at it. A few steps further they noted down on paper the objects they had caught. The son could soon write down forty objects. This training was kept up till an astounding ability was acquired. On the occasion of one performance the son gave the titles of more than a dozen books in another room, with the order of arrangement on their shelves. He had seen them in a single glance as he passed rapidly through the library.

Women as observers.

There are many women who have unintentionally educated themselves to a high degree of ability in quick observation. It can be safely asserted of many a one of them that, seeing another woman pass by in a carriage at full speed, she will have had time to analyze her toilet from her bonnet to her shoes, and be able to describe not only the fashion and quality of the stuffs, but also say if the lace be real or only machine made. It is said that, when passing on the street, eight women out of ten will turn around to see what the other one wears. I have often wondered at the two who did not turn around, but the reason is clear—they did not need to.

Quick observation in the schoolroom.

Innumerable exercises in quick and accurate observation can be used in direct assistance to the regular work of the schoolroom. The spelling of words can be learned by quick glances; the outline and parts of a country can be taught in greater and greater detail by successive quick exercises; a problem in mental arithmetic is to be grasped with only a momentary presentation of it; an object is to be drawn from an instantaneous glimpse; etc., etc. Indeed, there is not a single school exercise that cannot be so taught as to train this ability. In fact, the children are naturally quicker than we suppose them to be; it is often the case that lessons of interest to the child are carefully presented in such a way as to actually teach him to be slow instead of quick.

The arm-chair psychology.

But watching is not sufficient for science. "Learn to labor and to wait." For several thousand years psychologists have been waiting and watching; it never occurred to them to labor also. Sitting at home in the arm chair is very pleasant but it is not the way to do business, and consequently psychology has been going backward.

What is the reason that we to-day do not know how to train a child's mind properly? what is the reason that

M B X O
Q R A G
F C W P
T E D L

Fig. 2. An Exercise in Quick Observation.

philology is nothing more than a history of word-changes without an attempt to explain the causes? what is the reason that ethics is not a science but a conglomeration of maxims? what is the reason—but, stop, I will express it all by asking, What is the reason that the mental sciences to-day are two hundred years behind the physical sciences? The answer is sharp and decisive: Because the science of mind itself, psychology, owing to the late introduction of experiment, has not achieved the development that it should have done.

The advent of experiment.

It is to the introduction of experiment that we owe our electric cars and lights, our bridges and tall buildings, our steam-power and factories, in fact, every particle of our modern civilization that depends on material goods. It is to the lack of experiment that we must attribute the medieval condition of the mental sciences.

Contrast between observation and experiment.

In ordinary observation we wait for things to happen in one way or another; possibly they never happen in just the circumstances most favorable for studying them. In an experiment we arrange the circumstances so that the thing will happen as we wish. How good is the memory of a certain child? We might wait a long time before he happened to perform some memory exercise that would exactly answer the question. Instead of this we experiment on him by giving him lines of figures, sets of syllables, words, etc., till we know in just what condition his memory is. Galilei would never have discovered the law of falling bodies if he had not made the experiment.

The fundamental law of experiment.

Vary only one circumstance at a time. If you wish to find how strong a child's memory is at different times of the day, you should not make the morning test with words and the next with figures. There might be a difference due to the change from words to figures, and you would suppose this difference to be due to the time of day.

Grades of experiment.

Experiments can be divided into three grades. (1) *Tests.* The test is the simplest form and is an answer to the question, Is something so or not so? The usual test on hypnotized persons is pricking them with a pin to see whether they feel or do not; by flashing a light we determine whether a person is blind or not. (2) *Qualitative experiments.* By these we aim to answer the question, What? In experiments on the emotions we ask what bodily processes change with them. Given a person who can see ; to determine what he can see we make experiments for color-blindness. (3) *Quantitative experiments.* How much? is the question we ask in this case. How small a difference can you detect? how many syllables can you remember? how sharp is your vision? This is the highest class of experiments ; they are scientific experiments in the full sense of the word.

The objection is sometimes made that experiments in thinking, feeling, etc., are physical and not mental. This confuses the means with the thing, the tools with the work done. The apparatus is physical, but your accuracy of judgment, your suggestibility, your power of will, are mental.

Quack experiments.

It is perhaps advisable here to warn my readers against the unjustifiable application of the term "experiment" to hypnotic exhibitions, to thought-transference follies, and to the so-called psychical research experiments. These amusements are as unrelated to scientific experiments as clairvoyant healing or faith-cure to the science of medicine.

CHAPTER II.

TIME AND ACTION.

WHAT is the difference between a bicycle rider and a locomotive? The human body closely resembles a complicated machine. A man is the counterpart of an engine ; food is shoveled into the mouth of the furnace and is oxidized, *i. e.*, burned, in the body, producing heat. In the engine the heat is turned into motion by means of steam; the steam pushes the piston, which moves a series of levers. In the body the same result is reached in another way—by muscular contraction; the muscles of the leg, for example, move the complicated system of levers formed by the bones.

A conundrum.

If a locomotive is turned loose on the track—a runaway, as the engineers call it—it keeps right on till the fuel is burned out or it meets with a smash ; if the steam is turned off, it will stand motionless where you leave it. Not so the man ; he acts of his own accord and you can never be sure of what he will do next. The man or the bicyclist cannot be a mere machine ; there must be a governing power corresponding to the engineer on the locomotive. This governing power in man is the mind; it is just this power concerning which we are to busy ourselves.

The solution.

After all, the bicyclist does not resemble the locomotive till an engineer is put on. The answer to our conundrum is, therefore, the engineer.

Among the many problems in the science of mind we

naturally turn first to that of willing an act. Why does your hand move? As long as you had no will to move it, it remained still; but when you willed to move it, it moved. It is the will to move which preceded the act of moving. There is evidently some relation between the will and the act.

Willing, and its result.

Raise your hand. Did the hand move *when* you willed it to do so? or was the hand a trifle behind time? Here we have at the outset a knotty problem which all the discussion of a dozen arm-chair psychologists could not solve. Apparently the will and the act occur at the same moment; but we have grown so distrustful of "apparently" and "evidently" that we must remain in doubt till the case is proven one way or the other by experiment.

Time of will and time of action.

Here we have an example of how all psychological progress is limited to the invention of experimental methods and apparatus. The question just stated can be answered because a method of experiment has been devised; when the answer comes, there is a further question which we at once ask, but this one cannot be answered till some one finds the means of experiment. Since the question as to *time* has been raised, before inquiring what the answer and the second question are, you must learn how to measure time.

Need of experiment.

For the purpose of measuring small intervals of time one of the most convenient methods is the graphic method. Being one of the most beautiful and accurate methods of experiment, it is extensively employed in physics, astronomy, physiology, and psychology.

A method of measuring time.

The first thing to be done is to set up a tuning-fork—not a little one, such as musicians carry in the pocket, but one a foot long, vibrating one hundred times a

The fork, the marker, and the drum.

second. By means of a battery and a magnet this fork is kept going of itself as long as we please. The prongs of the fork move up and down one hundred times a second. Every time the lower prong moves downward, a point on the end dips into a cup of mercury, whereby an electric circuit is closed. This electric circuit passes through a little instrument called a time-marker, which makes a light pointer move back and forth also one hundred times a second. The point of the time-marker rests on a surface of smoked paper on a cylindrical drum. The smoked paper is prepared by stretching ordinary glazed paper around the drum and holding a smoky gas or benzine flame under it. A soft black surface is thus obtained, in which the point of the marker scratches a line as the drum is turned.

How they work.

Fig. 3. Apparatus for Recording Time.

When the time-marker is not connected with the fork, the point draws a straight line as the drum turns; but as soon as connection is made, it vibrates and draws a wavy line. Fig. 4 shows how the marker makes waves. To preserve the record, *i. e.*, to keep the smoke from

The record and its preservation.

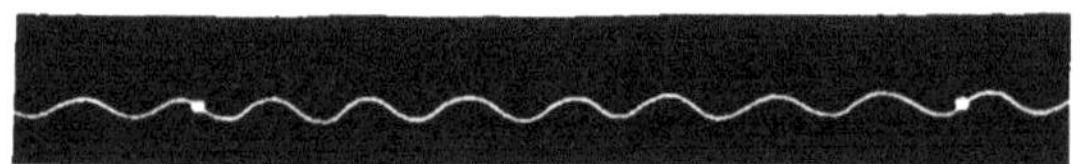

Fig. 4. A Specimen Record.

rubbing off, the paper is cut from the drum, run through a varnish, and dried, the result being what might well be called a study in black and white.

Reading the record.

Now, if the point of the marker moves back and forth just one hundred times a second, each complete wave must mean $\frac{1}{100}$ of a second. Consequently, if a dot be placed on the line at the moment I move my finger and another at the moment I move my foot, as is illustrated in Fig. 4, I can tell just how much time elapsed between the two movements by counting the waves and the fraction of a wave. Thus the two dots are distant by seven whole waves and five tenths of a wave extra; the time is, therefore, 7½ hundredths of a second, or 0.075^s.

Thousandths and hundredths.

In making careful records in the laboratory it is needful to count in thousandths of a second, but there is so much uncertainty about the last figure that in the final statement of results it is not only unnecessary to state the thousandths, but it is also misleading on account of the false degree of accuracy implied. We will therefore use hundredths of a second to count by. In order to save the multitude of o's and decimal points let us introduce the sign Σ to indicate hundredths of a second, just as ° indicates degrees. We will call the sign "sigma." Thus instead of 0.04^s we write 4^{Σ}.

The two dots.

We can always tell the time consumed if we can get the two dots. But how do we put dots on the line when things occur? That is just the difficult point; because we can find no method of making a dot at the moment of willing, we cannot tell just when the willing occurs. We have, however, found a way of making dots when most acts are performed.

Suppose we wish to make a dot when a finger is moved. The finger is placed on the button of a special telegraph key, so arranged that the slightest movement of the finger breaks an electric circuit. This electric circuit runs through a large coil of wire which makes a spark whenever the circuit is broken. Two wires run from this spark-coil, one to the drum and the other to a metal point resting on the smoked paper. Whenever a spark is made, it jumps through the paper, scattering the smoke and making a white dot. In Fig. 4 the metallic point was the time-marker itself. It is evident that every time we move the finger a dot is made.

Getting a dot by finger movement.

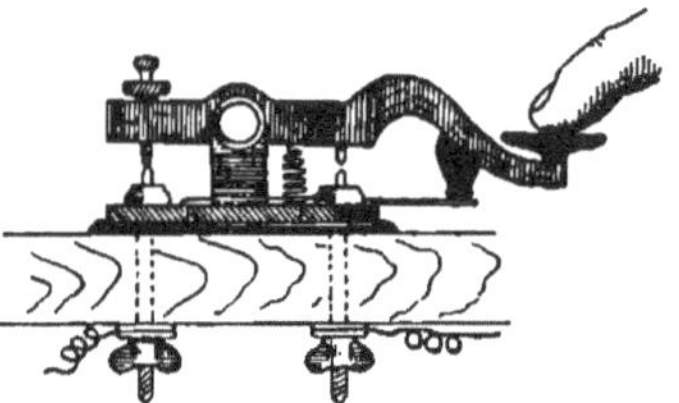

Fig. 5. Ready for a Record.

We wish, now, to find out if, when we will to move

The problem of the two simultaneous acts.

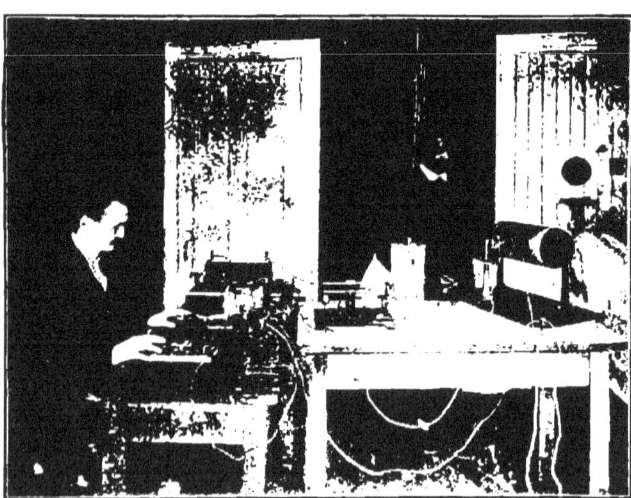

Fig. 6. Measuring the Simultaneity in Actions of a Piano-player.

the two corresponding fingers of the two hands at the same moment, they really do move as intended or if one is behind the other. To do this we

Its solution.

must have two keys, two spark-coils, and two metal points, one each side of the time-line. The plan of this arrangement is shown in Fig. 6. When the fingers move, two sparks fly through the paper and two white dots are made. Do they occur at the same moment? A specimen record is shown in Fig. 7.

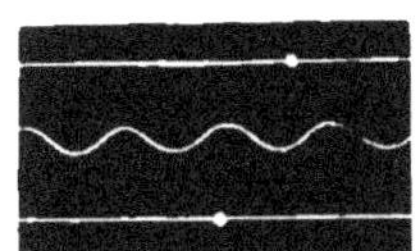

Fig. 7. Result of the Experiment in Fig. 6. The right hand (upper dot) is 0.005 of a second behind the left (lower dot).

Thus the will to move both hands at the same time results in moving the two at different times. A careful investigation shows that sometimes the right precedes,

Results differ from the intention.

sometimes the left, in irregular order. The difference frequently amounts to 1Σ and in a condition of fatigue may reach 5Σ. The difference may seem small. But, for example, the ear is very sharp and there are people in the world who, intending to strike the keys of a piano simultaneously, generally hit one slightly behind the other with a difference sufficient to be heard. Instead of playing music as written, such persons play, for example,

Practical applications.

There might be an educational value in using this method with many persons who cannot move two parts of the body at nearly the same time. Various exercises used in preparing speakers and actors, *e. g.*, simultaneous movements of head and hand, could be readily recorded.

Danger of overlooking small differences.

It is sufficient for practical purposes that the difference in time should not be noticeable. This is cared for by the instructor. It is to be remembered, however, that differences which the instructor may not notice will nevertheless be noticed by many of the audience. For example, the error of simultaneity in piano-playing might readily be great enough to produce a disagreeable impression on a large part of the audience and yet be so small as to have escaped the teacher's correction.

Although such means of testing simultaneity would be desirable for every piano-player, it is, of course, impracticable to provide smoked drums, spark-coils, etc., for general use. We must wait till some ingenious mechanic invents a hand arrangement to place directly on the piano keys.

The act occurs after the will to act.

We have thus answered our question. Since, when we will to move two hands at the same time, the actual movements occur at different times, therefore on each occasion the act of at least one hand is later than the will. As there is no reason to suppose that there is any radical difference between the two hands, it would be unjustifiable to draw any other conclusion than that the act is always behind the will.

How much?

And now for the second question. How much is the act behind the will? For the correct answer we must wait until the experimenter can find it.

This question is such an important one that we are forced to make a guess at it merely in order to get along with people. The time must be—as we can judge from our own experience—less than 1^s, or 100^σ; and, since one second is not worth noticing in ordinary matters of life, we can neglect the time entirely. This does not hold good in extreme cases. With persons influenced by curare the act does not follow àt all; in some diseases the act is much behind time. But for many practical purposes the act can be considered as occurring at the moment of willing. Such a case is that of rapidly repeated movements, which we will now consider.

Rapidity in tapping.

A large portion of the community depends on the rapidity with which it can will and execute certain acts. One of the elements of a good telegrapher is his accuracy and rapidity of tapping.

The experiments on tapping are most accurately made by the spark method we have just described. The finger is placed on a telegraph key, as shown in Fig. 5. The person is told to tap as rapidly as he can. Series of sparks fly off the end of the point of the time-marker in Fig. 3. On counting up the records we obtain the number of hundredths of a second for each tap. A good average rate is 15Σ per tap, or nearly seven taps to the second.

Fastest records.

The fastest tapping recorded is given as follows:

Middle finger	8Σ.
Hand	7Σ.
Tongue	7Σ.
Jaw	11Σ.
Foot	11Σ.

(Σ = 0.01 second.)

Fatigue.

The rapidity of tapping decreases with fatigue. Fig. 8 represents the results of a continuous series of taps, the lower the line the faster the tap; the straight horizontal line corresponds to a tap-time of 15Σ and the short checks on this line mark off the seconds. At first the tapping is rather irregular, but it is on the whole very rapid, one tap-time being only 11Σ. The tapping soon

Fig. 8. Influence of Fatigue on Tapping-time.

becomes steadier and remains rapid for about seventeen seconds. After that it is somewhat slower and more irregular, owing probably to fatigue.

Influence of mental work and disturbance.

The mental condition has a most powerful influence on the rapidity of tapping. Excitement makes the tapping more rapid. The influence of distraction of attention is shown in Fig. 9. This figure has the same meaning as Fig. 8. Adding 214 and 23 produced a marked slowness in tapping; so did the mental labor of multiplying 14 by 5. It takes some effort for an ordinary

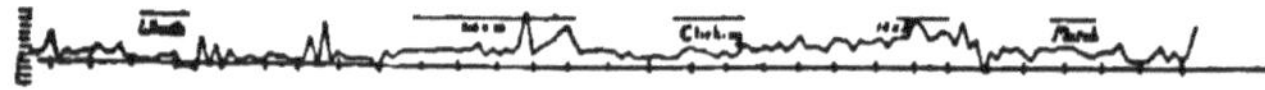

Fig. 9. Influence of Mental Activity on Tapping-time.

man to perform these calculations, and the mental work of association seemed to leave less energy for the work of will. The thought suggests itself of the possibility of measuring the amount of work involved in various school exercises by the influence on tapping.

The figure seems to show that momentary distractions not involving any work, such as whistling, clicking the tongue, or lighting a match, do not change the rapidity. They do, however, improve the *regularity;* the curve is smoother. It is a noteworthy fact in all our mental life that the less attention we pay to an act, the more regular it is.

Influence of the time of day.

The rapidity of tapping varies with the time of day. The averages of six weeks of work give the following results: at 8 a. m. the time required for making 300 taps is 37.8^s; at 10 a. m., 35.5^s; at 12 m., 34.6^s; at 2 p. m., 35.5^s; at 4 p. m., 33.5^s; at 6 p. m., 35.1^s.

Influence of habit.

It is noticeable that these results correspond to the habits of the previous two years of the person experimented upon; these years were spent in public school work with a daily program beginning at 8 a. m. and closing at 4 p. m., with an hour and a half intermission at noon.

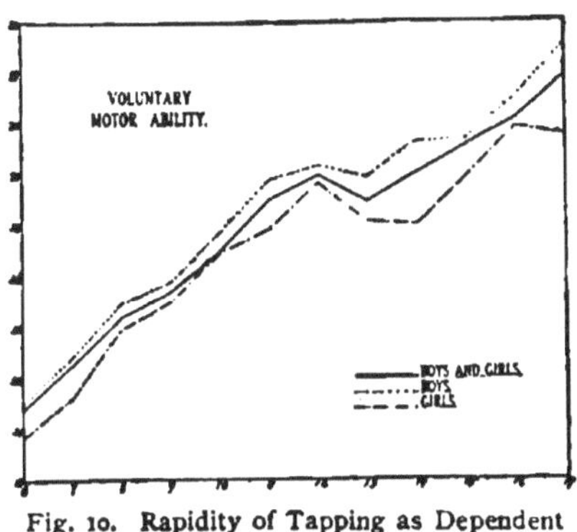

Fig. 10. Rapidity of Tapping as Dependent on Age.

Influence of age.

The rapidity of action increases steadily with age. Measurements of tapping-time on one hundred New

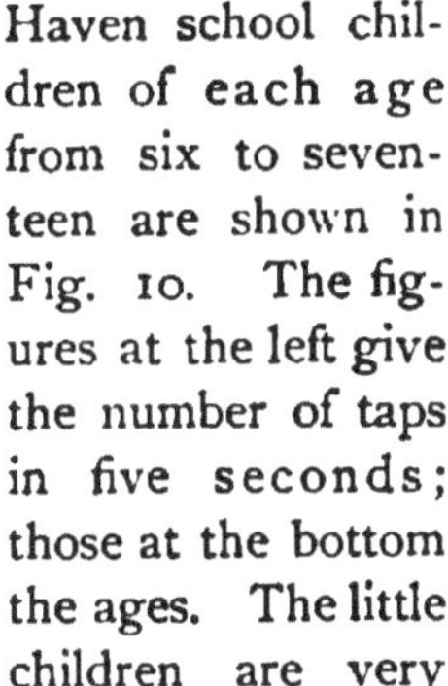

Haven school children of each age from six to seventeen are shown in Fig. 10. The figures at the left give the number of taps in five seconds; those at the bottom the ages. The little children are very slow; the boys at each age tap much faster than the girls.

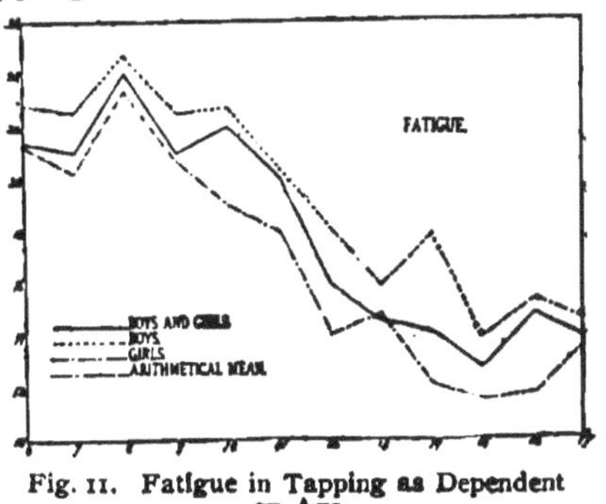

Fig. 11. Fatigue in Tapping as Dependent on Age.

Fatigue.

In these experiments the children continued tapping after the five seconds. After tapping thirty-five seconds longer a record was again taken. The difference between the two sets of records tells how much the child lost owing to fatigue. The results are shown in Fig. 11. The figures on the left give the percentage of loss; those at the bottom the ages. Thus, at six years of age the boys lost $\frac{23}{100}$, or 23 per cent of the original number of taps.

The amount of fatigue was greatest at eight years and decreased with advancing age. It is very remarkable that without exception of a single age the girls were less fatigued than the boys. A comparison of the two figures suggests a conclusion as to the impetuosity of the boyish character.

CHAPTER III.

REACTION-TIME.

A series of reactions.

WHEN you signal to the car conductor to stop, he reacts by pulling the bell-strap, the driver reacts to the sound of the bell by pulling the reins, and the horses react by coming to a rest. By reaction, then, we will understand action in response to a signal. The time between the moment of the signal and the moment of the act is known as the reaction-time.

The chain-reaction.

Is there any such time? Quick as thought—that must be pretty quick. Let a number of persons stand in Indian file as if about to march; each one places his right hand on the head (or shoulder) of the person in front.

Fig. 12. A Series of Reactions.

Bend the file around till a complete circle is formed with every right hand on the head of the one in front. One of this file we will call the experimenter; in his left hand he holds a watch—preferably a stop-watch. All the rest close their eyes. The instruction is given: Whenever you feel a sudden pressure from the hand on your head, you must immediately press the head of the person in front. When the second-hand of the watch is at the beginning of a minute, the experimenter presses the head of the one in front, he presses that of the next in front, and so on. The pressure thus passes all around the group and finally comes back to the experimenter. At the moment he feels the pressure he notes how many seconds have passed. Suppose there were ten persons in the circle and the

Fig. 13. Chain-reaction.

watch has gone three seconds; then three seconds is the time required for ten acts in response to a signal. The average time for one reaction is obtained by dividing the number of seconds by the number of persons; thus, in this case the reaction-time would be $\frac{3}{10}$ second, or 0.3^s.

Almost all the main experiments in reaction-time and thinking-time can be illustrated in this way by a group of ten or more persons. Some of the most interesting I will indicate after describing the more accurate methods.

Others can be readily devised by any one ingenious at games.

Time is required for reaction.

It takes time, then, to react. A hundred years ago people did not know this. And thereby hangs a tale.

The story of the astronomer's assistant.

Astronomers have to record the moment of the passage of a star across lines in a telescope. In 1795 the British astronomer royal found that his assistant, working with another telescope at the same time, was making his records too late by half a second. Later on, this difference amounted to 0.8^s. This difference was large enough to seriously disturb the calculations, so the poor fellow lost his place for the sake of eight tenths of a second.

The story of the two astronomers.

Many years later two famous astronomers were observing the stars together and recording their passages across the telescope. Strange to say, one was steadily behind the other. Now it would not do to make accusations against a noted astronomer ; this set people to thinking. One of the astronomers went to a third astronomer and again there was disagreement. Finally, after more experience, astronomers in general reached the conclusion that everybody disagreed with everybody else. Moreover, men who disagreed in one way at one time would be likely to disagree differently at another time ; so that a man did not even agree with himself. As this was evidently not the fault of the star, the conclusion was finally reached that each person had a peculiar error of his own. This was called by the queer name, "personal equation." The British astronomer, who did not suspect that he himself might be incorrect, was perhaps no nearer right than his assistant. At any rate, the actual time of passing of the star differed from the recorded time.

They all disagree.

Not the fault of the star.

Americans are noted for asking, "How much?" Science is, in this respect, merely concentrated Americanism; it always asks, "How much?"* It is not sufficient to know that we are always behind time; let us make a systematic inquiry of how much time we lose. A miss is as good as a mile, but it is a very interesting thing to know just how bad the miss is. Therefore we will get to work systematically to measure just how much time we lose in acting to a signal.

How much time is required?

Systematic work proposed.

Even the best of us is inattentive. So to be rid of all distraction the person experimented upon is put in a queer room, called the "isolated room," whose thick walls and double doors keep out all sound and light. When a person locks himself in, he has no communication with the outside world except by telephone.

The isolated room.

Fig. 14. In the Reaction-room.

All the sights and sounds can be shut out, all disturbances of touch can be made small by comfortable

How it feels.

* There are so-called "qualitative sciences" that have no methods of measurement or statistics. These are the demireps of the scientific world with whom we must put up because we haven't more respectable members of society to take their places.

chairs, but, alas ! we have let in a sad source of disturbance, namely, the person himself. Let me describe what I hear and see in the silence and darkness of the room. My clothes creak, scrape, and rustle with every breath ; the muscles of the cheeks and eyelids rumble ; if I happen to move my teeth, the noise seems terrific. I hear a loud and terrible roaring in the head ; of course, I know it is merely the noise of the blood rushing through the arteries of the ears (what you hear when you place a shell to the ear), but I can readily imagine that I possess an antiquated clock-work and that, when I think, I can hear the wheels go 'round. As for the light—great waves of lavender-colored light sweep down in succession all over the darkness in front ; beautiful blue rings with purple centers grow and grow and burst, only to start over again in different colors. The physiologist tells us that these are merely effects of chemical processes going on in the eye ; and, indeed, everybody sees these things when he closes his eyes at night. But that does not help us to get darkness.

A source of disturbance.

Be it as it may, the results are far more reliable than those obtained in ordinary laboratories and under ordinary conditions of distraction by the rattle of the streets, the banging of college clocks, the buzzing of machinery, and the commotion of students.

More reliable results.

It must not be thought that the invention of this room is an imputation on my part against the attentive powers of humanity. It is simply a fact, to which we must all plead guilty, that we cannot pay attention amid the bustle and roar of life around us. It is easy to imagine what a boon an office on this plan would be to a busy banker or a newspaper editor.

An explanation.

Having put the person in the isolated room with

nothing but electrical connection to the apparatus in the other rooms, we will begin by asking how long he requires to react to a sound. We will use the graphic method, as illustrated in Fig. 3, with the addition of the spark method explained in Chapter II. Let us be modern and send the sound by telephone. A multiple key, in which electrical currents can be combined in forty-one different ways, is so arranged that by pressing it a sound is sent through a telephone and at exactly the same moment a spark is made on the drum.

Reaction to sound.

The various arrangements for making the experiments have been developed to a high grade of accuracy and convenience. In the recording room the smoked drum stands on the table, the electric fork is in front, the multiple key in the middle. The multitude of wires and accessory apparatus seems to make a hopeless chaos, but they are all carefully arranged for convenience and accuracy.

Great accuracy obtained.

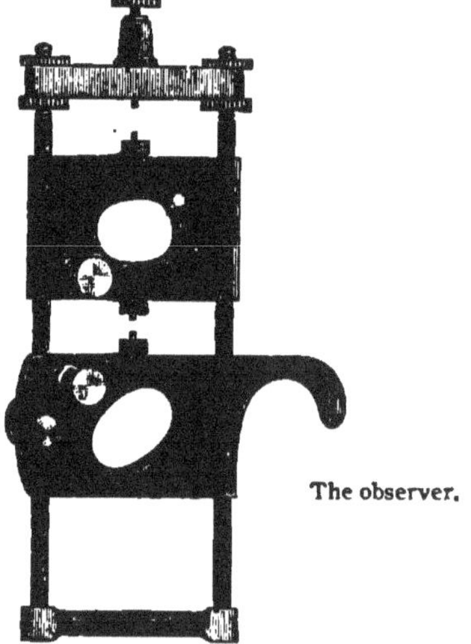

Fig. 15. Reaction-key.

The person in the isolated room—let us call him the observer, for short—sits comfortably with the telephone at his ear and with a curious electric reaction-key (Fig. 15) in his hand (the ordinary telegraph key has proved itself too awkward and fatiguing). The forefinger is placed in the hole of the smaller, or movable, slide, and the thumb is placed in the hole or against the hook of the lower, or adjustable, slide. Flexible wires lead to the post at the top and to the movable slide. The hand is placed in any convenient position, and the

The observer.

The reaction-key.

thumb and finger are held apart. The slightest movement of the finger makes a spark on the drum.

The experiment.

When the multiple key is pressed, the telephone circuit is closed and whatever sound is sent through the transmitter is then heard by the person experimented upon. At exactly the same instant a record is made on the time-line on the drum. The moment the sound is

Fig. 16. Reaction to Sound.

heard by the person experimented upon, he moves the finger in the reaction-key ; thus a second record is made on the time-line. A record similar to that of Fig. 4 is obtained ; the number of waves, however, will depend on the particular person, the particular sound, etc.

The average results.

For noises the reaction-time is a trifle shorter than for tones. For example, a person who reacts to a noise in 11^{Σ} will take perhaps 15^{Σ} for a tone. Even the whistle of a locomotive is not so conducive to a quick jump by the passengers on the platform as a sudden escape of steam.

Reaction-time at the start of a race.

A particular case of reaction to sound is found in starting a race. In short-distance, or sprint, racing the time required for the reaction is a very important factor. The starter's pistol is fired and the racers are off, but the man with a very short reaction-time will have gained a respectable fraction of a second over the other. To measure this reaction-time an electric contact is put on the end of the starter's pistol. The arrangement is shown in Fig. 17. The firing of the pistol causes the wing to fly back and break an electric circuit, thus making a record. A runner's key of the kind shown in Fig. 18 is attached to the runner by a thread. The start of the runner jerks and breaks the thread; this moves the lever and makes another record.

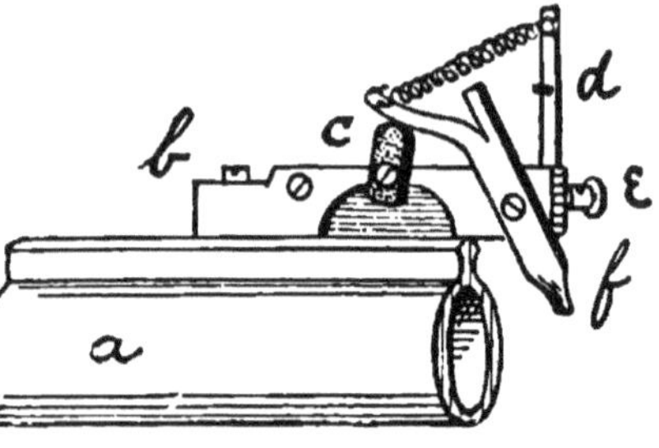

Fig. 17. The Pistol-key.

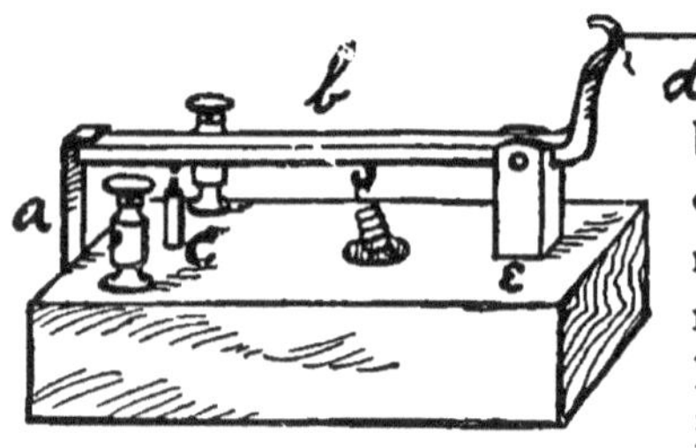

Fig. 18. The Runner's Key.

Some results.

Although I have never had time to carry out an extended series of experiments on racers, the experiments made have shown a few facts. The first point noticeable is that long-distance runners are very much

Fig. 19. Measuring a Runner's Reaction-time.

slower than sprint runners who practice quick starting; this shows that the reaction-time can be reduced by practice. The reaction-time seems to be longer where the whole body has to be started than where only a finger is moved ; the mass to be moved thus seems to have an influence on the time. In some races the pistol has gone off and the photograph has been made of the runners before they have reacted.

Reaction-time to touch.

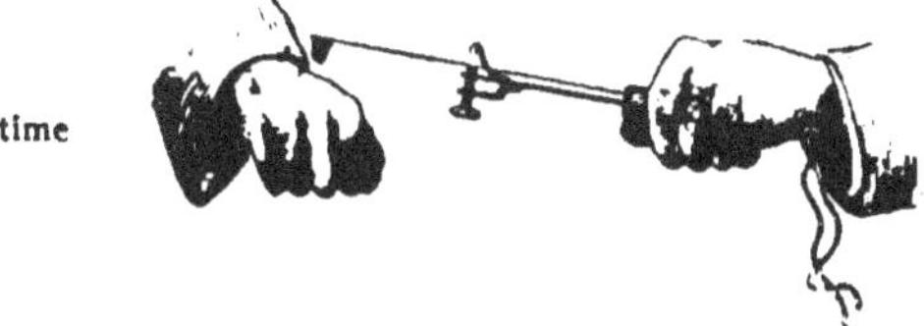

Fig. 20. The Touch-key.

The reaction-time to touch can be found by using an instrument shown in Fig. 20. The flexible conductors carrying the current pass through the screws of this stimulator and then through the reaction-key. The person experimented upon closes his eyes. Some one takes the stimulator and touches him, whereupon he reacts by moving his finger as before. The stimulator makes a record on the drum and so does the key.

Relation to intensity.

It can be laid down as a general law that a weak touch

Fig. 21. Reaction to Touch.

is answered by a slower reaction than a moderately strong one. As the touch becomes stronger the reaction-time decreases, but when it becomes very strong the time is again lengthened. The moral is this: if you want time to recover after dealing a blow, hit your antagonist very hard or almost not at all.

To experiment on the reaction-time for temperature-sensations a metal ball is screwed on the touch-stimulator in place of the rubber tip. The ball is heated or cooled as desired.

Fig. 22. A Reaction to Cold.

Reaction-time to hot and cold.

The reaction-time for cold is somewhat shorter than that for hot, and both are longer than for touch. For example, the figures for one experimenter are: touch, 11Σ; cold, 12Σ; hot, 13Σ.

The reaction-time to light is found by using an electric

Reaction-time to light.

flash instead of the telephone. The intensity of the light has a very great influence. A very weak light might give 33^{σ}, while a strong one would give 20^{σ} for the same person.

The photographer's application.

This interval renders it possible for the photographer to get perfectly natural flash-light pictures. The flash goes off, the picture is taken, and all is again dark in a couple hundredths of a second. But such a small time is quicker than reaction-time and so the whole is done before the persons can move.

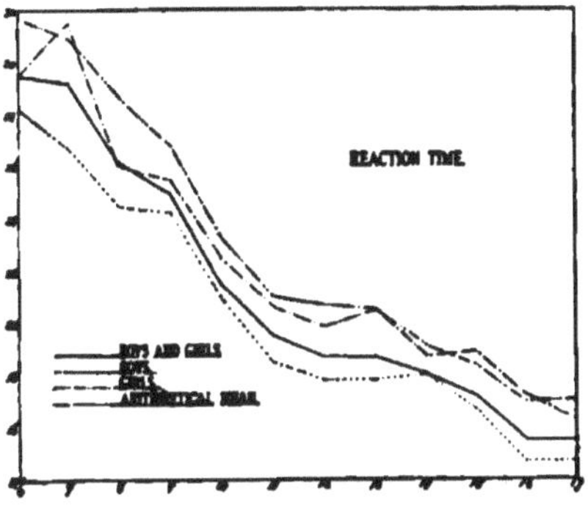

Fig. 23. Reaction-time Decreases with Age.

Reaction-time in children.

Children become steadily quicker as they grow older. The results of the New Haven measurements are shown in Fig. 23. The figures at the left indicate the number of hundredths of a second required for reaction to sight ; those at the bottom the ages. The topmost line in the figure relates to another matter. Boys are much quicker than girls at each age—that is, in simple reaction ; how they compare in quickness of thought will be told in the following chapter.

CHAPTER IV.

THINKING-TIME.

Time of thought.

THE simple reaction-time has led to a method of measuring the time of thought. One of the fundamental processes of thought is recognition. To determine the time of recognition the subject reacts on one occasion just as quickly as he can, without waiting to notice what he is reacting to. In popular phrase, he hits back without waiting to know what struck him. Recognition cannot be said to be present. On the next occasion he fully recognizes what he hears, sees, or feels before he reacts. The difference in time between these two cases gives the recognition-time. Properly speaking, the former reaction would be the true, simple reaction, but this distinction is often overlooked and both kinds are then lumped together.

Recognition-time.

Experiments on one subject gave the following recognition-times: for a color, 30Σ; for a letter, 54Σ; for a short word, 52Σ.

These times refer to experiments where the person is ready and eager to recognize the object. How long it would take to recognize an object unwillingly, *e. g.*, a tradesman by an English snob, has not yet been determined.

Recognition of complex objects.

A single figure, such as a triangle or a square, is recognized as quickly as a simple color. We can grasp enough of a triangle to recognize it without attending to details; a three-cornered figure is as simple as a color when nothing but its corneredness is noticed. A single letter takes the same time as a short word. The total impression of a well-known object is so familiar to us that we need no more divide it into its parts in order to distinguish it than we do in the case of a simple color. In reading we do not divide the word into its letters, we grasp the word as a whole by a single thought.

Recognition of letters.

The various letters of the alphabet require different times for recognition. There are slight differences for letters of different sets of type; they vary from 0.6^{σ} (0.006^{s}) to 5^{σ} (0.05^{s}). The following sets of letters are arranged in the order of time required.

Good.	Fair.	Poor.
m w d g v y j p	k f b l i g h r x t	o v a n e s c z
m w p q v y k b	**d j r l o n i g h u**	**a t f s x z c e**
d p q m y k n w	o g v x h b j l i a	t u z r s c f e

German letters.

A German requires 1^{σ} to 2^{σ} more time to recognize a letter of his antiquated alphabet, for example, **𝔴**, than to recognize a letter in the Latin type, **w**. But in reading words no more time is required to recognize the word in either case. The twists and tails of the old letters cause a loss of time in recognizing a single letter, but in grasping the words only the main features receive attention anyway.

Discrimination-time.

Another of the fundamental processes of thought is discrimination: Is it white or black, loud or weak, hot or cold? In all cases it takes time to decide.

Discrimination-time for tones.

Suppose that the person on whom the experiment is made is to discriminate between two different tones. In addition to the arrangement described on page 43, we require two tuning-forks of different pitch. The sound is sent by telephone as before. The person is told not to react till he has recognized which tone he hears. Sometimes one tone is sent, sometimes the other. If we determine the person's reaction-time for a single tone, where he knows that only one tone is used, and also the reaction-time with discrimination between two tones, we are justified in subtracting the former from the latter, and calling the result the "discrimination-time" for two tones. In a similar manner the discrimination-time for three, four, or more tones can be measured.

For sight.

The discrimination-time for sight can be very prettily illustrated. The Geissler tubes are filled with different gases so that when an electric current is sent through them they show different colors. An induction-coil (or spark-coil) is fitted up so that the current can be sent through any one at pleasure. To get the simple reaction-time one tube alone, *e. g.*, a red tube, is used, the time between the flash and the reaction being measured as before. Then two, three, etc., are used, just as mentioned for tones. Ordinary times for discrimination can be represented by the following specimens: for two objects, $8\underline{\Sigma}$; for three, $14\underline{\Sigma}$; for four or five, $15\underline{\Sigma}$.

Choice-time.

The next element of thought-life to be considered is choice. How shall we determine the choice-time? The Geissler tubes can be very conveniently used for this purpose. The subject places his five fingers on a five-knobbed telegraph key. When he sees the red light he is to press his thumb; when he sees the yellow he is to press his forefinger; and so on. There are thus five objects for discrimination and five movements between which to choose. Of course the time is much lengthened. If

Results.

we know the discrimination-time and reaction-time for five colors, we can subtract these from the total time with choice, thus getting the choice-time for five. It is evident that the choice-time for two, three, four, six, or more objects can be similarly found. One subject gave a choice-time of 8Σ for two fingers, with steady increase up to 40Σ for ten fingers.

After some practice with the same fingers for the same colors, the act of choice gradually falls out and the movement becomes associated to the color. The extra time still remaining might be considered as a kind of association-time for movements. The association-time in the usual meaning is measured differently.

Time of discrimination and choice in chain-reaction.

The time of discrimination and choice combined can be obtained from a group of persons without any other apparatus than a watch. The persons of the group stand in a ring, as shown in Fig. 13, each with his hand on his neighbor's head. In the first place, the simple reaction-time is measured by giving the head a slight push and sending the push all around the circle, as described on page 39. "Next time," says the experimenter, "each of you will receive a slight push on the head either forward or backward. You are to send the push along in the same way."

The experiment is made three or four times, sometimes with a forward push, sometimes with a backward one. Each person, not knowing which he is to receive, will be obliged to discriminate and then choose the appropriate movement of the hand. By subtracting the simple reaction-time from this last result, the time for discrimination and choice for two things is obtained.

Then the experiment is repeated with three movements: right, left, or forward. Then with four: right, left, forward, or backward. The time will be found to grow longer as the number increases.

The time of association of ideas, which is what is usually meant by association-time, is best measured by calling out words or showing pictures to some one who is to tell what he associates to each. For example, I call out "house" and you say "street." **Association-time.**

A peculiar mouth-key is placed before the transmitter in the recording room and a somewhat similar one in the experiment room. The experimenter shouts some word, *e. g.*, "glass." This causes the thin plate in the mouth-key to rattle and make a spark record on the drum. At the same moment the subject hears the word in the telephone at his ear. He shouts back what he first thinks of, *e. g.*, "water." This makes a similar record. The total time between the two records less the discrimination-time and choice-time will give the association-time. **The mouth-key.**

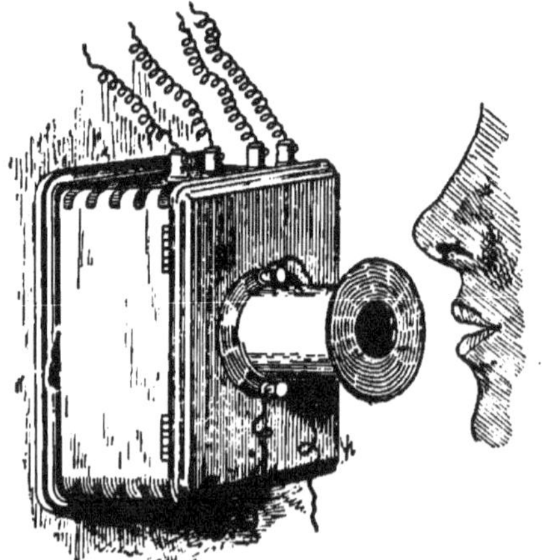

Fig. 24. Reaction by Voice; or, the Voice-key.

The associations may be of various kinds. In "free" association, the subject thinks of whatever he pleases. **Free association.**

The time for free association can be put in the neighborhood of 80Σ.

Forced association. In a "forced" association the subject is allowed to associate only objects bearing certain relations to the object presented. Thus, whenever he hears the name of a country he must name one of its cities. In such a case he has a moderate range of association. In a strictly forced association there is no freedom. Thus, whenever the name of a person is mentioned, his native land must be associated.

1 × 8 = 8
2 × 8 = 16
3 × 8 = 24
4 × 8 = 32
5 × 8 = 40
6 × 8 = 48
7 × 8 = 56
8 × 8 = 64
9 × 8 = 72
10 × 8 = 80
11 × 8 = 88
12 × 8 = 96

Fig. 25. Strictly Forced Associations.

Results. As specimen results we can give the following association-times: translation from one's own language to one a trifle less familiar, 15Σ to 30Σ; giving the succeeding month of the year, 25Σ to 30Σ; simple addition of two figures, 12Σ to 22Σ; simple multiplication of two figures, 25Σ to 35Σ.

A particular form of association is found in the logical

judgment. In fact, many of the forced associations are really abbreviated logical judgments. Suppose it to be required to associate the whole when a part is given, *e. g.*, given "root," associated "tree"; this is simply a practical abbreviation of "a root is a part of a tree." More difficult cases can be devised. It holds good as a general rule that in actual thinking the forms of logical thought become forced associations.

Logical judgment.

All our acts are complications of thinking-times, simple reaction-times, and action-times. Study of mental and muscular time for practical purposes has been made in only a few cases.

Mental and muscular time in fencing.

The visit of several expert swordsmen to Yale furnished the opportunity for some experiments on their rapidity in some of the fundamental movements of fencing.

First experiment.

The first experiment included a determination of the simple reaction-time and of the time of muscular movement. The fencer stood ready to lunge, with the point of the foil resting to one side against a metal disk. A flexible conducting cord, fastened to the handle of the foil, hung in a loop from the back of the neck. A large metal disk was placed directly in front of the fencer at a distance of 75^{cm}. Just above this disk was a flag held on a foil by an operator standing behind it. A movement of the flag was the signal upon which the lunge was executed.

The spark method of recording was so arranged that the primary circuit passed through the electric switch, a spark-coil, the flexible conducting cords, the foils, and either one of the two disks. Every make and break of this circuit made a spark record on the drum. As long as the foils rested against the disks the current was closed. The movement of the flag-foil broke the circuit for an instant, making a record of the moment of signal. The first movement of the fencer's foil broke the circuit again at the small disk, making a record of the moment of reaction. The striking of the foil against the large disk made a third record. The time

Fig. 26. Measuring Mental and Muscular Time in Fencing.

between the first and second records gave the simple reaction-time ; that between the second and third gave the time of movement through the given distance. About ten experiments were made on each person.

Second experiment.

In the second experiment the flag-foil was moved in various directions. The point of the foil rested against the small disk. The movement in any way of the flag was the signal for a corresponding movement of the foil. Acts of discrimination and choice were thus introduced into the reaction-time. The movement of the foils gave records as before. The time required can be called the reaction-time with discrimination and choice. About ten experiments were made on each person.

The persons experimented upon consisted of Dr. Graeme Hammond, Dr. Echverria, Dr. P. F. O'Connor, and Mr. Shaw (all expert amateur fencers), A. Jacobi, master of arms of the New York Athletic Club, Prof. Ladd, formerly practiced in fencing, and Prof. Williams, with no knowledge of fencing.

The results were:

Results.

1. Simple reaction-time: Echverria, 17Σ; Williams, 19Σ; Hammond, 19Σ; Ladd, 23Σ; Jacobi, 23Σ; Shaw, 23Σ; O'Connor, 26Σ.

2. Time of muscular movement involved in the lunge through 75^{cm}: Jacobi, 27Σ; O'Connor, 29Σ; Echverria, 31Σ; Shaw, 32Σ; Hammond, 32Σ; Ladd, 52Σ; Williams, 57Σ.

3. Reaction-time with discrimination: Hammond, 22Σ; Ladd, 24Σ; Williams, 25Σ; Jacobi, 29Σ; Echverria, 30Σ; Shaw, 36Σ; O'Connor, 36Σ.

Conclusions.

The experiments probably derive their chief value as calling attention to the experimental study of the psychological elements involved in games, sports, gymnastics, and all sorts of athletic work. Without experimenting on large numbers of fencers and others, I would not attempt to make any quantitative comparisons between the two. The following qualitative conclusions seem, however, to be fully justified.

1. It is possible to analyze fencing movements into their mental and bodily elements, and to measure them.

2. The average fencer is not quicker in simple reaction (where a few mental elements are involved) than a

trained scientist, and neither class shows an excessive rapidity.

3. When once the mind is made up to execute a movement, fencers are far quicker in the actual execution. In rough figures, it takes them only half as long as the average individual.

4. As the mental process becomes more complicated, the time required by the average fencer is greater than that required by a trained scientist. The shortest time of all, however, is that of Dr. Hammond, whose mental quickness has probably been developed in some other way.

Fencing develops muscular but not mental quickness.

5. The general conclusion seems to be that fencing does not develop mental quickness more than scientific pursuits, but it does develop to a high degree the rapidity of executing movements. It would be important to determine if this holds good of the other sports and exercises, or if some of them are especially adapted to training mental quickness.

Mental and muscular time in arm movements.

In order to study the quickness of movements of the arm we use the apparatus shown in Fig. 27. A horizontal brass bar carries on it three adjustable blocks, A, B, and C. The block A has a flag which may be suddenly

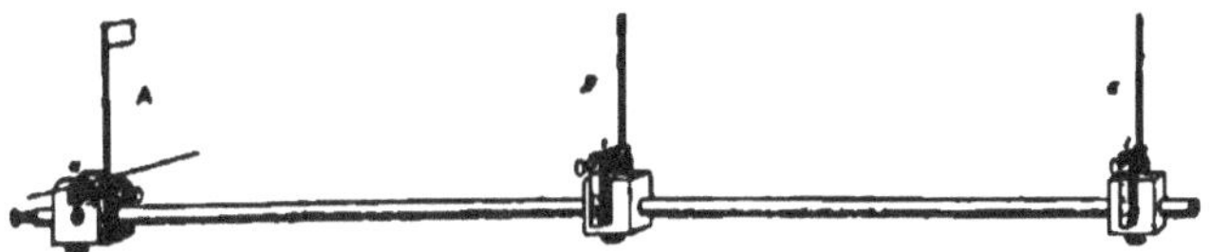

Fig. 27. Apparatus for Measuring Rapidity of Thought and Action.

jerked to one side by a thread. The other blocks have light bamboo sticks projecting upward. The whole arrangement as used on a pugilist is shown in Fig. 28. The boxer takes his position and places his fist just behind the stick at C. At the moment the flag moves he is to strike straight out. The apparatus is connected with the spark-coil and the recording drum.

Time of reaction and action of a pugilist.

The flag is jerked; this makes a spark on the time-line. The boxer strikes, knocking down both sticks. Each stick makes a spark also. We thus have three sparks on the time-line. The time between the first and the second gives the simple reaction of the boxer; that

Fig. 28. Measuring how Rapidly a Pugilist Thinks and Acts.

between the second and the third gives the time required for the fist to travel the distance between the two sticks.

Time of discrimination and choice.

The boxer is next told that the flag will be jerked to the right or left in irregular order and he is to punch only when it goes to the left. He is thus obliged to discriminate and choose. Sparks are obtained as before, but the time between the first two dots is longer because two extra mental acts are included.

It is possible to use not only men, women, children,

athletes, pugilists, and others as animals for experiment; we can also use dogs and cats. One of my pupils has contrived a similar arrangement for measuring how fast a dog thinks. The general plan is shown in Fig. 29. The results cannot yet be made public, as it is the rule of the laboratory to let the experimenter speak first.

How fast a dog thinks.

In the New Haven experiments the school children

Fig. 29. Measuring how Fast a Dog Thinks.

were required to distinguish between two colors, reacting to blue and not to red. This involved the mental processes of discrimination and choice, in addition to simple reaction. The results are shown in Fig. 30. The figures at the bottom indicate the ages; those at the left give the number of hundredths of a second required for reaction with discrimination and choice. The topmost line does not concern us here.

Time of discrimination and choice as depending on age.

The time required decreases with age. On the whole, the boys and girls are equally quick, the differences generally being too small to be worth noticing. It might be suggested that, since boys are quicker in simple reaction, they must take a longer time for mere discrimination and choice in order to give equal totals. The figures seem to indicate that for the more involved mental processes the girls are quicker, but I hesitate to admit such a libel on my own sex.

Girls and boys compared.

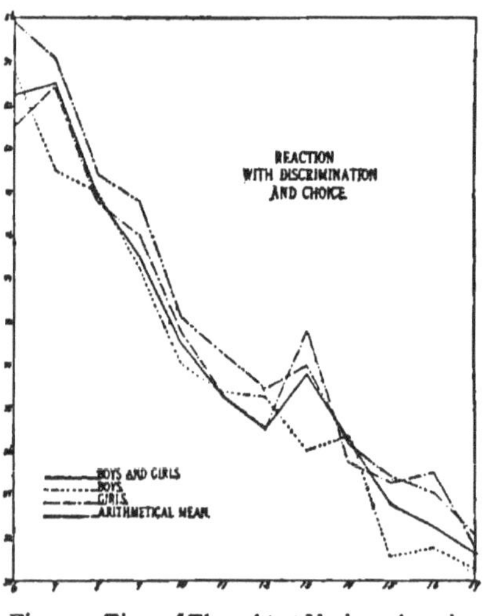

Fig. 30. Time of Thought at Various Ages in School Children.

Irving said that Americans worship only one god, the Almighty Dollar. He was wrong; there is a mightier one, the Moloch of Time.

Time, the great independent variable, is the only force over which we can gain no control. Man can annihilate space and fight power with power, but—tick, tick, tick—the little watch counts off the seconds, not one of which can be hindered from coming or be recalled when past.

Mighty time.

Time is the most precious of commodities. No one wants a six-hour train to Boston when a five-hour train is at hand. Slow horses to the engine give the fire a fatal opportunity. Battles have been gained by the quickness of the cavalry. Death may readily be carried at the sword's point of a quick antagonist.

Value of time.

Time is money. Rapid thought and quick action sometimes make all the difference between success and failure. Every thought we think, every act we perform, takes time. A man who can think and act in one half the time that another man can, will accumulate mental or material capital twice as fast. If we could think twice as fast as we do, we would live twice as long, although we would live only the same number of years. Country people think more slowly than city people ; the uneducated more slowly than the educated. In general the Americans are very rapid thinkers. To-day the mental processes of the mass of the people go at a much more rapid rate than they did a few centuries ago. The mind has been educated by our whole civilization to act more rapidly. To-day our thoughts travel like trolley-cars.

Civilization has decreased the time of thought.

The difference between the sluggish Englishman of medieval times and the quick Yankee of to-day is delightfully told in Mark Twain's "King Arthur." If it were possible to take a man of two centuries ago and bring him into the laboratory, the results obtained from experiments upon him would be entirely different from those obtained from one of the students of to-day. The reactions of the student would be much more rapid, especially the complicated ones. A great deal in the education of children is to reduce their reaction-times. When the country boy first comes into the schoolroom everything he does takes him a very much longer time than when he has been there for a while, especially any complicated act. Arithmetic, for example, is simply a matter of the association of a set of ideas. We give just so much time to do an example. When that time is over the pencil must be put down, the slate dropped. The child who is slow is at great disadvantage. Education in arithmetic, especially mental arithmetic, has for its object mainly the reducing of the time in associating ideas ; say one half toward producing a firm memory of the associations and the other half in making them more rapid.

Education of rapidity.

Rapidity in movement and thought is a part of our education to which we must pay some attention. Mental rapidity is increased by repetition, provided the repetition does not continue long enough to bring opposing forces into play. For example, the oftener we repeat a poem from memory, the more rapidly it can be done, provided we do not become fatigued.

Method of training.

To increase the rapidity of the act it is not sufficient to simply repeat it without fatigue ; unless there is present a conscious or unconscious determination to change the time of the process, there is no reason for expecting it to change. The first requisite for increase in rapidity is thus a desire for such an increase.

Illustration from arithmetic.

Let it be required to increase the rapidity with which a child performs his arithmetical associations. If allowed to do his sums in any time he pleased to take, constant practice might not cause the slightest change in rapidity. If, however, he were stimulated by hearing the pencils of his comrades, by seeing them finish before him, or by the general influence of a bright, sunny day, he might do his work more rapidly, although he had not had the slightest intention of doing so or perhaps even the knowledge that he had done so. Such influences might be called unconscious motives. By a conscious motive we would mean a definite intention of getting the sum done more quickly. Our general experience in life justifies us in believing that a conscious motive is more efficient than an unconscious one.

Increased rapidity in language-lessons.

The manner in which rapidity of thought is increased by practice in learning a language has been made the subject of experiment. Ten boys were taken from each class of a high school and were asked to read rapidly the first hundred words of a Latin book. The number of seconds that they required is shown in the following list:

Class 10,	average age	9,	average time	262^s
" 9,	"	" 11,	"	" 135^s
" 8,	"	" 12,	"	" 100^s
" 7,	"	" 13,	"	" 89^s
" 6,	"	" 14,	"	" 79^s
" 5,	"	" 15,	"	" 57^s
" 4,	"	" 16,	"	" 54^s
" 3,	"	" 18,	"	" 49^s
" 2,	"	" 19,	"	" 48^s
" 1,	"	" 22,	"	" 43^s

The lowest class knew nothing about Latin, the rest had begun it in Class 9.

When the same children were tested with their native language the results were successively 72^s, 55^s, 43^s, 37^s, 39^s, 28^s, 27^s, 26^s, 25^s, 23^s. There was a similar gain.

Where the gain lies.

Was the gain due to general gain in mental rapidity? One hundred papers of five familiar colors were shown and each child was required to name them. The average times were 83^s, 66^s, 79^s, 66^s, 63^s, 56^s, 63^s, 63^s, 54^s. There had been a general gain in quickness but not nearly so great a gain as for the words. A study of the blunders made by the children showed that in the next to the lowest class there was a very slight tendency to grasp the Latin letters as words; they blundered occasionally by reading a similar word for the correct one. In the succeeding and higher classes this mistaking of words became steadily more frequent; they had been trained to grasp larger groups as single things and in this manner to save time in discrimination.

Language in the printery.

This same ability to grasp a greater quantity of matter by means of characteristic marks, whereby the details can be supplied as needed, is what makes the difference between English and Latin in the composing room. While setting up English the compositor works by the em, that is, by quantity ; while setting up foreign words he works by the hour, as such work is very slow.

Highly educated rapidity.

How far we can push the education of rapidity in all the elements that make up thinking-time, reaction-time, and action-time can be seen in the records for rapidity of telegraphing and typewriting.

By careful estimate it has been found that in general press matter the average number of letters per word is five, and that the average number of vibrations of the key in the formation of the telegraphic characters is five to each letter. Thus it is seen that there are twenty-five vibrations of the key in the formation of each word. Now, were it possible for an operator to transmit sixty words per minute, he would make one word, or five letters, per second, being twenty-five vibrations of the key per second.

Rapidity of thought and action by the telegrapher.

When we consider that the telegraphic alphabet is made up of dots and dashes and spaces of various lengths, and that these almost incredibly rapid vibrations must be so clear and clean cut as to be easily read by the ear, we can form an approximate idea of the wonder of such an achievement. The most rapid manipulator in the country has reached a speed of fifty-four words per minute, which is about $23\frac{3}{4}$ vibrations of the key per second. Expert typo-telegraphers can receive and record his transmissions with ease.

Championship record.

By use of the Phillips system of code words an expert transmitter and typo-telegrapher can handle press matter at the rate of from sixty-five to seventy words per minute. One noted telegrapher claims to have read by sound from automatic transmission up to seventy words per minute, which is in the neighborhood of thirty vibrations of the instrument per second. To do this the ear must be long and carefully trained to be able to distinguish and translate into words and sentences the sounds coming to it in such rapid vibrations. It would be impossible to read from a transmission much beyond this speed.

CHAPTER V.

STEADINESS AND CONTROL.

STEADINESS of action may be steadiness of position or steadiness of movement. In position the impulses to the various muscles are so arranged that the member or the body remains still. In movement the impulses are varied in power in such a way that a change occurs.

Steadiness of action.

In studying action, voluntary or involuntary, we need

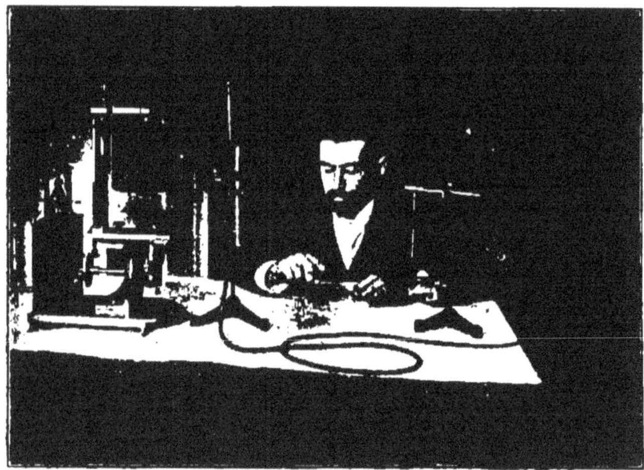

Fig. 31. Taking a Record of Steadiness.

to have some method of recording every part of the act. This is found in the principle of air transmission.

Air transmission.

In investigations of the steadiness of position we gen-

The capsules. erally make use of a pair of capsules. Each one consists of a little metal dish covered with thin rubber. From one dish a tube leads to the other. A very light lever is placed above each dish; the lever carries a light plate which rests on the rubber top. If one of the levers is moved downward, as by the hand in Fig. 31, the rubber will be pressed in and the air will be slightly pressed out through the tube. The pressure will pass along the tube to the other capsule, where it will bulge the rubber top and will make the other lever move upward. When the lever is released, the spring will draw it back, the air will be drawn in, and the other lever will move downward.

The record. To make a record, a fine metal point is attached to the second lever and is made to write on a surface of smoke. A metal cylinder is covered with paper and is then smoked in a gas flame, as previously described. The most frequently used cylinder for slow movements

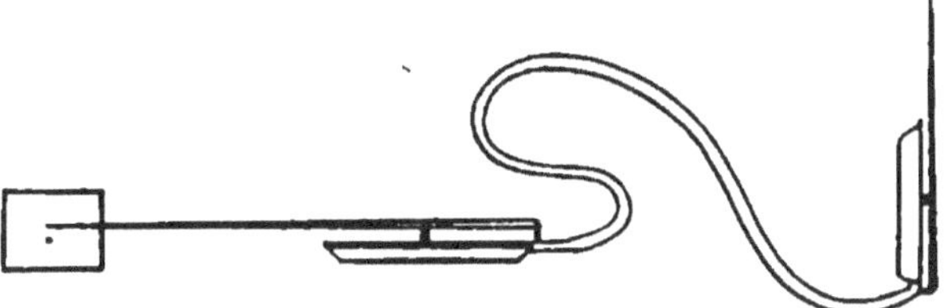

Fig. 32. Arrangement of Capsules for Steadiness under Guidance of the Eye.

is a clock-work drum of the kind shown in Fig. 31. The fine point of the second, or recording, lever is made to touch the surface of the smoked paper. The point then writes a picture (upside down) of the movement imparted to the end of the receiving lever.

Steadiness of the arm. Let us now take some particular problem, such as the steadiness of the arm, guided by the eye. The arrangement can be that shown in Fig 32. Every shake of the arm will be transmitted to the recording point and will be scratched in the smoke on the drum. Under guid-

Fig. 33. A Record of Steadiness.

ance of the eye the position can be kept the same; whether the steadiness increases or decreases remains to be determined. The lever of the receiving capsule is made very long. Its point is held by the finger opposite a dot on a card. It is impossible to keep the point opposite the dot; there is constant shaking.

Under guidance of the eye.

A specimen record is given in Fig. 33. During the time between the two vertical strokes the attention was disturbed by some one walking around the room.

Let us study steadiness in a concrete case, say in hold-

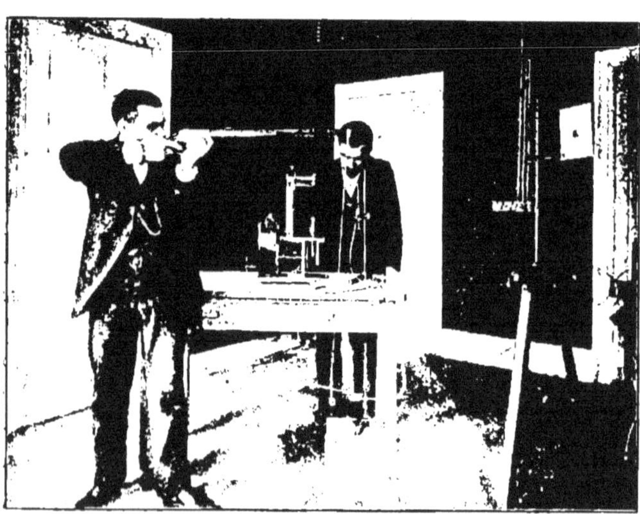

Fig. 34. Recording a Sportsman's Unsteadiness.

A sportsman's steadiness.

ing a gun. The sportsman takes his position, standing, with gun aimed at the target. A thread hangs down from the gun with a small sinker at the end to keep it stretched. The thread is given one turn around the arm of a receiving capsule, as seen in the figure.

Steadiness in standing.

The method is a wonderfully convenient one and can be applied to a study of almost every position taken by the body. By placing the arm of the receiving capsule on the head, as shown in Fig. 35, a record of the fluctuation in height can be made.

Fig. 35. Steadiness in Standing.

Trembling of the hand.

Persons inclined to loss of control over their muscles often show symptoms in early life. It is well to test the steadiness of the hand in children. A very convenient method of studying the trembling of the hand is shown in Fig. 36. The capsule is connected with the recording capsule as before.

Trembling of the tongue.

By a tongue-capsule, as in Fig. 37, we are able to tell how still the tongue can be held. If it should be proven possible by the trembling of the tongue in childhood to foretell which persons would become talkative in later life, precautionary measures might be taken.

The most interesting fact about these experiments in

Will as opposed to execution.

steadiness is that the *will* is to have a steady position but the *execution* is defective. As the will is exerted the steadiness of position is increased. This is sometimes so marked as to be visible to the eye directly. I have seen the scalpel tremble in a surgeon's hand so that a serious

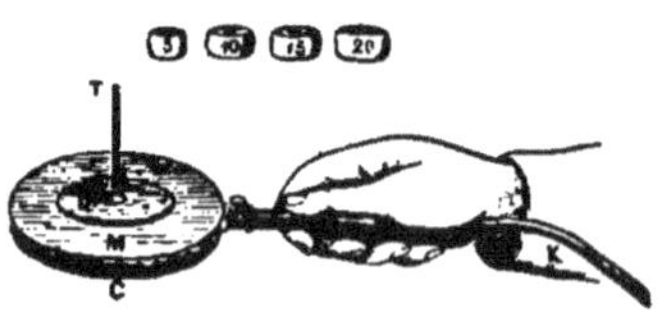

Fig. 36. Studying the Trembling of the Hand.

Fig. 37. Studying the Steadiness of the Tongue.

accident appeared inevitable ; yet when the supreme moment came the hand guided the knife with admirable steadiness.

Steadiness of movements.

Proceeding from steadiness for position we come to the question of steadiness of movements. Owing to the difficulties of apparatus, this subject has not received so much study as the other. There are, however, several very practical and simple experiments that can be made.

Drawing a straight line.

In free-hand drawing it is frequently desirable to make a straight line. The line as actually made is always more or less irregular. I once proposed the following problem : Suppose we desire to draw a line 100^{mm} long, what is the most accurate method of making it straight? As it was most important to know the facts for school children, the experiments were performed on ten boys.

Method of experiment.

The boys all sat at their desks in just the same positions. A sheet of paper seven inches long by four inches wide was placed before each. In the middle of the sheet were two dots 100^{mm} apart, lengthwise of the paper. At a given signal each boy drew a straight line between the dots. Afterwards a ruler was laid on each sheet so that its edge cut the dots. With a pair of dividers the greatest deviation of the line drawn from the true straight line was found.

Results.

In the first sets the boys sat squarely in front of the desk, holding the pencils in the usual way, grasped near the middle. The line was drawn with a single movement of the pencil, without going over it a second time or erasing. The first line drawn was horizontal, *i. e.*, parallel to the front surface of the body. On the second set of papers the line drawn was vertical, the other conditions remaining the same. In the third set the line was 45° to the right, in the fourth 45° to the left. The positions of these lines can be thus shown :

0° ⟶ 270° ↓ 45° ↗ 325° ↘

The facing position proved to be more favorable for horizontal and vertical lines than for inclined lines. The right-side position is also more favorable for horizontal and vertical than for 45° and 325°. This is what we might expect as a result of the law that the eye moves more easily upward, downward, right, and left (*i. e.*, vertically and horizontally), than in intermediate positions.

Explanation.

In drawing horizontal lines and 325° lines the right-side position is more favorable than the facing position ; for the others facing is preferable. This is perhaps to be explained by the fact that the forearm swings around the elbow in a curve which in order to produce a straight line must be compensated by a backward and forward movement of the upper arm around the shoulder. In the facing position, with the paper directly in front, the forearm touches the body at the start and the hand is bent at the wrist. As the arm moves, it becomes freer and a more natural position is assumed. This change in the manner of carrying the arm would tend to introduce uncertainty into its movements. With the arm raised upon the desk in the right-side position it is brought clear of the body, and the line can be executed in one sweep. In drawing the 45° line the arm is just as free in the facing as in the right-side position and we find little difference in the results. In drawing the vertical line we would naturally expect much greater accuracy when the motion is a simple forward or backward movement of the arm around the shoulder, as in the facing position, than when the arm has to undergo complicated adjustment with the elbow raised. Why there should be a difference with the 325° line it seems impossible to say. Both positions, facing and right side, are on the whole equally favorable for accuracy.

Advantage of various methods.

Holding the pencil far from the point is in general the most accurate method; near the point is as accurate as the middle grip. With the pencil far from the point the line is drawn with a smaller movement of the hand, which would give a better result than a larger movement requiring adjustments from elbow and shoulder. For horizontal lines the far grip is the most accurate; for 45° the same is true; for vertical lines the middle and the far grips are the same, the near grip is unfavorable; for the 325° line the near grip is the best, the far grip is next, the middle grip is very unfavorable. That the 325° line forms an exception to the advantages of the far grip and is much less regular than the others, is evidently connected with the awkward contraction of the fingers in this direction.

Influence of practice.

Can steadiness be increased by practice? This problem can be answered in respect to the hand. The arrangement for measuring steadiness has been made very simple, involving no capsules or drums. It consists of a flat block of hard rubber supported vertically by a rod. On the face of the block is a strip of brass in which there are five hard rubber circles, 1^{mm}, 2^{mm}, 3^{mm}, 4^{mm}, and 5^{mm} in diameter. The edges of the circles are flush with the brass. The object is to touch the rubber circle with the metal point at the end of a stick by a single steady movement. Sufficient unsteadiness of the hand will cause the point to touch the metal. With the same circle the steadiness of the hand can be considered to be directly proportional to the percentage of successful trials. To indicate when the metal point strikes the plate instead of the circle, an electric current can be sent from one pole of a battery through an electric bell to a binding-post connected with the metal plate, and from the other pole through a flexible conductor to the metal point. Any contact of the point with the plate will cause the bell to ring.

The steadiness-gauge.

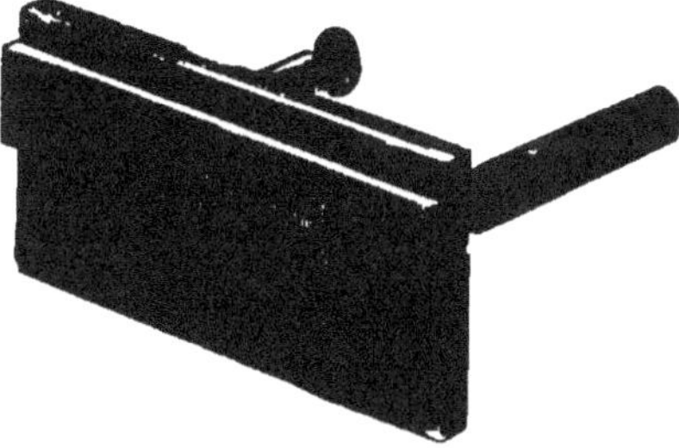

Fig. 38. Steadiness-gauge.

Making the experiment.

In making the experiment the plate is set up in front of the person experimented upon. The pointer is grasped in the middle like a lead pencil; the forearm is rested on a cushion at the edge of the table and the trial is made by a single steady movement under guidance of the eye (Fig. 39).

A series of experiments on the subject of steadiness was lately carried out in my laboratory. The first set consisted of twenty experiments with the left hand ; the result was fifty per cent of successful trials. Immediately thereafter twenty experiments were made with the right

Results.

Fig. 39. Measuring Steadiness and Attention.

hand, with the result of sixty per cent of successful trials. On the following day and on each successive day, two hundred experiments were taken with the right hand, the same conditions in regard to time, bodily condition, and position in making the experiments being maintained as far as possible. The percentage of successful trials ran as follows : 61, 64, 65, 75, 74, 75, 82, 79, 78, 88. The increase in accuracy is represented in the curve in Fig. 40.

Cross-education.

On the tenth day the left hand was tested with twenty experiments as before, with seventy-five per cent of successful trials, thus showing an increase of twenty per cent without practice in the time during which the right hand had gained as shown by the figures above. This curious process I have ventured to call "cross-education."

The question of the possibility of gaining in steadiness

by practice is thus definitely settled. We find also that the left hand gains by practice of the right.

Kindergarten work.

Let us notice in passing how much these experiments resemble the cork-work, bead-work, perforating, and weaving of the kindergarten, and the sewing of higher classes.

Steadiness in singing.

The pitch of a tone sung from the throat depends on the tightness with which the vocal cords are stretched by the muscles of the larynx. If a singer can keep these

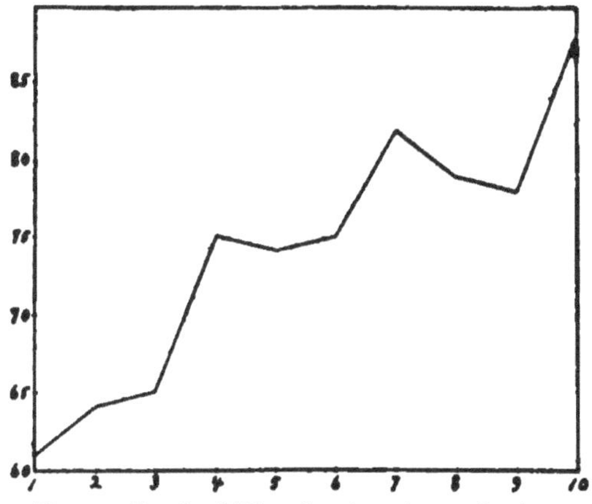

Fig. 40. Result of Educating Attention to the Arm.

muscles steady in position, the tone remains the same; if he allows them to change ever so little the tone changes.

Method of experiment.

A means for studying the accuracy of singing a tone, and keeping it, is found in the gas-capsule and mirror-tuning-fork. The gas-capsule consists of a little box (Fig. 41) divided into two parts by a thin rubber membrane. A gas pipe leads to one part and a small burner is attached. The person sings into the other part. Every vibration of the voice shakes the membrane and makes the little flame bob up and down too rapidly to be seen. This flame is placed in front of a tuning-fork having a little mirror on one end. The tuning-fork is set

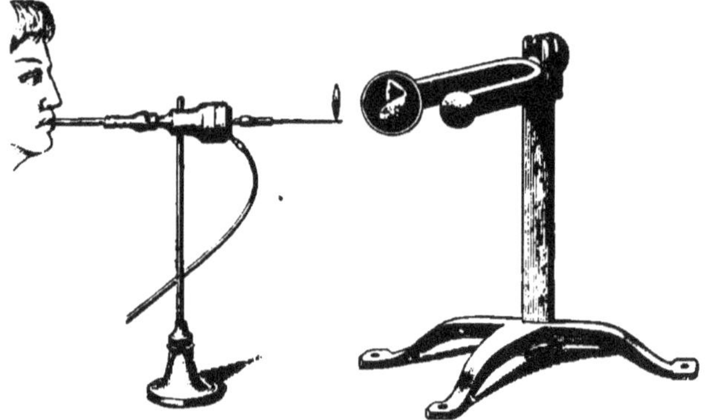

Fig. 41. Testing Steadiness in Singing. The Unison.

going and the person sings the same tone. A curved flame with a single point appears in the mirror.

Effect of unsteadiness.

Any inaccuracy or change in the pitch of the singing makes the picture rotate in the mirror. If it rotates in the way the flame points, the person sings too low; if backwards, then too high. If the singer is only a trifle wrong, the rotation is slow; a poor singer makes the picture fly around at all sorts of speeds.

The apparatus can do more than this. Suppose the fork is tuned to middle C. Then let the person sing the tones indicated by the quarter-notes, the half-note

Results for various intervals of pitch.

indicating the tone of the fork. When the unison is sung, a flame with a single point is seen. When the octave is sung, a double-pointed flame appears (Fig.

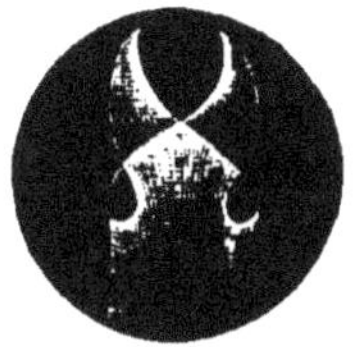

Fig. 42. Singing the Octave.

Fig. 43. Singing the Duodecime.

Fig. 44. Singing the Fifth.

42). For the duodecime we get three points (Fig. 43); for the double octave four points. These points seem to be upright, but for musical intervals, such as the fifth, the pointed flames are twisted together. For the fifth

Fig. 45. Singing the Fourth.

Fig. 46. Singing the Third.

we see three points twisted as in Fig. 44; for the fourth we get Fig. 45; for the third, Fig. 46.

When these intervals are properly sung the flames appear sharp and steady; any inaccuracy causes rotation. The apparatus thus tells directly how steadily the singer maintains his pitch.

CHAPTER VI.

POWER AND WILL.

WHAT is the relation between the force of will and the force of action? What was the difference between Samson slaying the lion and Samson shorn of his locks? Was the will the same in both cases? At one moment we will to grip the pencil tightly, at the next loosely; in the first case it cannot be taken from us, in the second it readily slips. Does the act correspond to the will?

Force of will and force of action.

We can, at least, measure one thing, namely, the force of the act. Numerous dynamometers—as the instruments thereto are called—have been invented. The simplest, possibly the best, form of a dynamometer is an ordinary spring-scale. An arrangement for studying the strength of pressure between thumb and forefinger is shown in Fig. 47. The iron frame carries a spring-scale of the appropriate strength. The thumb is placed on the cork and the finger on the hook of the scale. When the two are pressed together the pointer on the scale shows the amount of force exerted. Spring-scales of various strength can be used. Dynamometers have been constructed for the

The dynamometer.

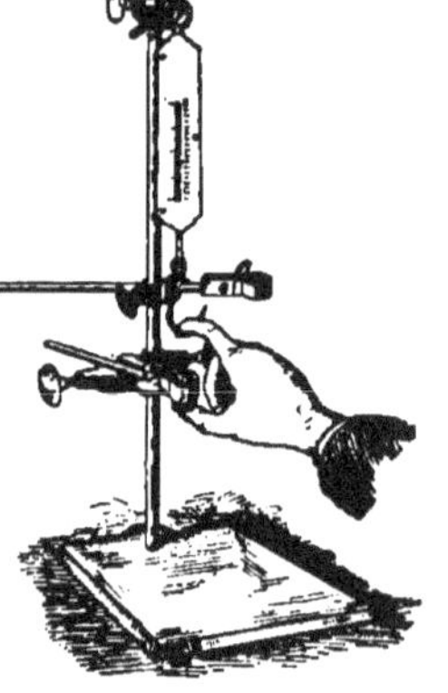

Fig. 47. Spring Dynamometer.

hand, foot, knee, extension of the two arms, lifting by the back, and so on. They are all, I think I can say, merely rough instruments for testing and have never been developed into scientific apparatus.

Accuracy in exerting power.

How accurately can we exert our force of will?

When on his return home Ulysses desired to punish the insolence of the beggar, Irus, by inflicting a severe blow, yet feared lest the well-known power of his arm would betray him if he put forth his whole strength, he deliberated on the amount of force to be employed,

> "Whether to strike him lifeless to the earth
> At once, or fell him with a measured blow,"

and decided to deal one which would only fracture the jaw. This was evidently a very fine regulation of the amount of exertion.

Use of the dynamometer.

As a matter of experiment let us determine the accuracy for the thumb-and-finger-grip with the dynamometer of Fig. 47. The swinging stop at the back is so fixed by the collar that when the stop extends across the frame, the hook strikes and hinders further movement. The movement is arranged to stop at, say, one pound or 500 grams.* The person is seated with the eyes closed. The stop is swung on and the pressure is exerted till the hook strikes. This is a pressure of 500 grams. The finger is released, the stop is turned back, and the experiment is repeated. As the person finds no hindrance, he stops when he thinks he is exerting the same force as before. The actual position of the pointer is read off and the error is noted. Suppose he stops at 495^g; he then makes an error of -5^g.

Results for the whole hand-grip.

Experiments with the whole squeeze of the hand indicate that, if on an average a man makes a mistake of

* In scientific work and in civilized countries the gram is the unit of weight.

twenty grams on a weight of 200 grams, he will make one of 30g on 400g, 40g on 800g, and 46g on 1600g. As the weight grows larger, the actual amount of the average error (or average mistake) increases; thus it is 20g on 200g and 46g on 1600g. But the proportion of the error is not increased or even the same, but is decreased; 20g is a much larger part of 200g than 46g of 1600g.

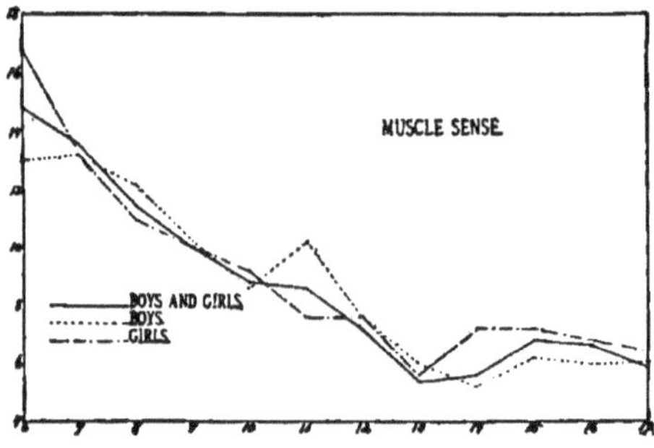

Fig. 48. Decrease of Inaccuracy of Weight-judgments in School Children of Successive Ages.

"Muscle sense."

A method of making similar experiments with the arm is to lift cylindrical weights between thumb and finger. The weights are sorted into two groups, those that appear the same as the standard used and those that appear different. The amount of difference that passes unnoticed gives an idea of the accuracy of the judgment. This is generally said to be a judgment by the "muscle sense."

Force of act and force of will.

Up to this point we have experimented on the force of the voluntary act and have said nothing about the relation between the force of will and the act itself. The force of the act we have measured in pounds or grams. Will, not being a mechanical process, cannot be measured by any physical force; it can be measured only in terms of will.

Two acts of equal force.

By making use of the same method of reasoning as in regard to time we can draw a general conclusion in regard to the force of act as dependent on the force of will. Suppose we will to pull with the two hands with just the same force ; do the results differ ?

To solve the problem we use two dynamometers. The person experimented upon squeezes with the right hand and the left hand with what he considers equal force. On looking at the scale the records are read and the difference noted. Here are some I have obtained. The + indicates that the right hand was stronger, the — the same for the left : each record is the average of ten experiments.

A. F., janitor,	— 8.5	ounces
M. S., woman,	— 7.4	"
W. S., girl, 3½ years	+ 0.5	"
" " "	+ 1.2	"

Neglect of small differences.

We see that although the will is apparently alike in both cases the acts are not. Nevertheless, since the difference is a small fraction of the force of act, we can say that within corresponding limits the force of will can be considered to be indicated by the force of act. For most of the points we are about to consider we can take any differences in the force of act as representing corresponding differences in force of will.

Cultivation of accuracy of force.

The importance of a cultivation of accuracy of force in making an effort is known to those who play ball, billiards, tennis, bagatelle, or quoits. The smith and the gold-worker must hit with just the right intensity. The proper force in the breathing movements is what the speaker and the singer have to learn.

Strongest possible effort.

The strongest possible effort has received more attention than the accuracy of effort. In using the dynamometer for these experiments the spring is given free play and the effort is made as strong as the person can make it.

The greatest power obtainable with a determined effort of will varies from the strength of Hercules to the feebleness of an invalid. The actual amount of force obtainable from various persons is a problem of anthropology, with which we are not concerned ; we shall use the force of action as a means of studying will power.

Dependence on mental condition.

The greatest possible effort depends on the general mental condition. The greatest possible effort is greater on the average among the intelligent Europeans than among the Africans or Malays. It is greater for intelligent mechanics than for common laborers who work exclusively, but unintelligently, with the hands. Intellectual excitement increases the power. A lecturer actually becomes a stronger man as he steps on the platform. A schoolboy hits harder when his rival is on the same playground. A bear's fear for the safety of her cubs might well be considered proportional to the number of pounds difference in the force of her blow. I venture to suggest that the difference between the greatest possible effort when alone and the greatest possible effort when before a roomful of ladies be used as the measure of a young man's vanity.

Effect of practice.

The amount of force corresponding to the greatest possible effort is increased by practice. It is incredible to me how in face of our general experience of gymnasium work some writers can assert that practice makes no change in the greatest possible effort. At any rate, in experiments made under my direction the change could be steadily traced day by day.

Cross-education in power.

Curiously enough, this increase of force is not confined to the particular act. In the experiments referred to, the greatest possible effort in gripping was made on the first day with the left hand singly and then with the right hand, ten times each. The records were: for the left, 15 pounds, for the right, 15 pounds. Thereafter, the *right* hand alone was practiced nearly every day for eleven days, while the left hand was not used. The right hand gained steadily day by day; on the twelfth day it recorded a grip of 25 pounds. The left hand recorded on the same day a grip of 21 pounds. Thus the left hand had gained six pounds, or more than one third, by practice of the other hand.

Physical exercise and will power.

A great deal has been said of the relation of physical exercise to will power. I think that what I have said sufficiently explains how we can use the force of act as an index of will power. It is unquestionable that gymnastic exercise increases the force of act. The conclusion seems clear; the force of will for those particular acts must be increased. It has often been noticed that an act will grow steadily stronger although not the slightest change can be seen in the muscle.

Strength and will.

Of course, I do not say that the developed muscle does not give a greater result for the same impulse than the undeveloped one; but I do claim that much of the increase or decrease of strength is due to a change in will power. For example, no one would say that Sandow, the strong man, has a more powerful will than anybody else. But Sandow's strength varies continually, and, although part of this variation may be due to changes in the muscles, a large portion is due to a change in force of will. When Sandow is weak, make him angry and note the result.

Changes in strength.

The power exerted varies according to what we hear, feel, or see. Music, colors, emotions, change our strength.

With the thumb-and-finger-grip the greatest pressure I can exert during silence is eight pounds. When some one plays the giants' motive from the Rheingold my

Influence of music.

grip shows 8¾ pounds. The slumber motive from the Walküre reduces the power to 7½ pounds. Let

me suggest to my readers that they rig up a simple dynamometer and keep it beside the piano in order to measure the stimulating power of each composition. The effect of martial music on soldiers is well known. The Marseillaise was a force in the French Revolution.

Plato on the character of Greek music.

Just how much of the inspiriting effect is due to the rhythm, the time, the melody, and the harmony, has not been determined. A very great deal depends on the pitch. Plato emphasizes the influence of the proper music on the formation of character. He goes no further than to specify the general scales in which music should be written. The high Lydian is plaintive, the Ionian and Lydian are soft and convivial, the Dorian is the music of courage, and the Phrygian of temperance. Aristotle agrees in general but considers the Phrygian music as exciting and orgiastic. It has long been supposed that

Probable explanation.

the difference among the scales was one of arrangement of the intervals within the octave, corresponding to the major and the minor, but the more recent opinion is that the difference is one of pitch. The Lydian is a tone to a tone and a half higher than the Phrygian, and the Dorian is a tone below the Phrygian. The Dorian is neither too high nor too low, and expresses a manly character.

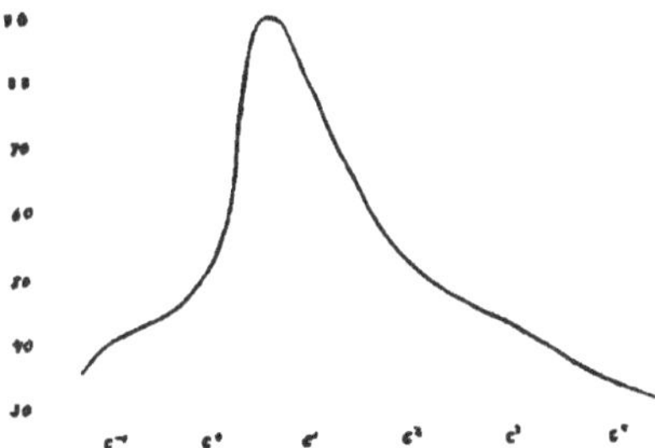

Fig. 49. Influence of Pitch on the Power of Grasp.

Influence of pitch.

It might be suggested that the special melodies associated with each scale may have had much to do with the case. Nevertheless it has been proven that the pitch itself has an effect on the greatest strength of grip. Fig. 49 shows the number of pounds for one person's hand-grip as the scale was run up on the piano.

Fig. 50. Dynamograph.

In order to indicate the pressure continuously the

dynamometer can be arranged in connection with the capsule of the graphic method. One such arrangement is shown in Fig. 50. As the hand squeezes the dynamometer the pointer on the drum moves to one side. Every fluctuation in the squeeze is shown, and when the smoked paper is taken from the drum and varnished we have a complete record. Such a tracing from an hysterical person squeezing as hard as possible is shown in Fig. 51. Dynamograph.

The power of squeeze is changed by various disturb- Influence of noises.

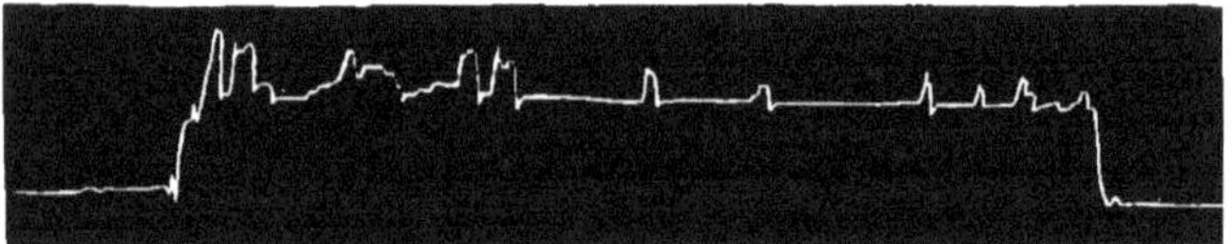

Fig. 51. Record of Strongest Grip of the Hand by an Hysterical Person.

ances. The sudden jerks in the line of Fig. 51 are the results of the ringing of a gong. The sudden increases in power occur each time when the gong is struck.

Successive single contractions can also be registered on

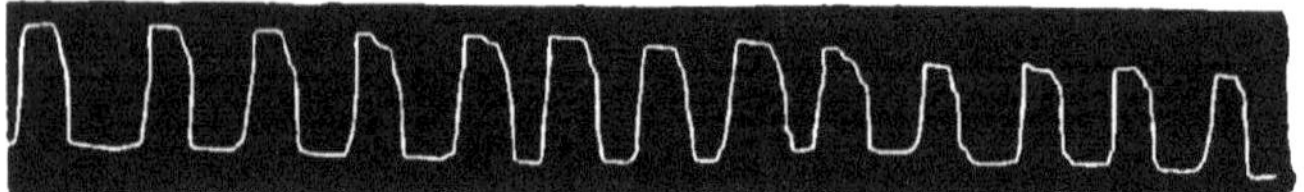

Fig. 52. Record of Successive Squeezes during Ringing of a Gong and during Silence.

the drum. Fig. 52 shows the successive squeezes of one person with the hand—first while a gong was being sounded, then in silence. The use of the gongs on the trolley-cars as a strengthening tonic might be suggested to physicians.

The colors also affect the squeeze with some persons, Influence of colors.

especially hysterical people. The strongest hand-squeeze in the case of one such subject is shown in Fig. 53.

Influence of smell.

This suggests a new principle in the selection of colors for the house, for uniforms, etc. We know the stimulating effect of the red flag of anarchy and the soothing influence of a dark blue sofa. A red bedquilt is a contradiction.

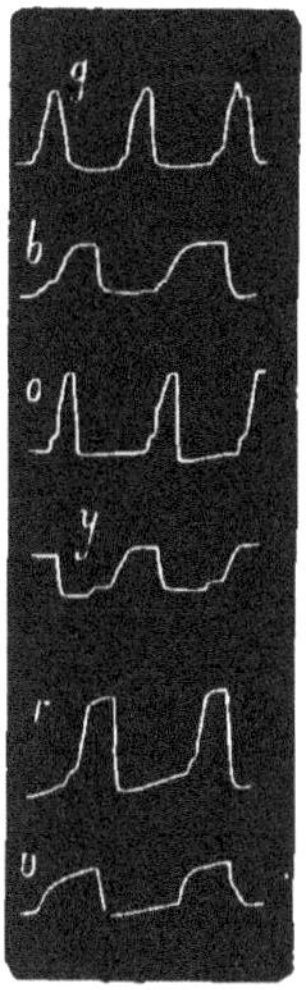

Fig. 53. Strongest Contractions while Looking at Different Colors: *g*, green; *b*, blue; *o*, orange; *y*, yellow; *r*, red; *v*, violet.

Tastes and smells have different effects. Fig. 54 shows the effect of musk on the power of a tired person. Tobacco has a stimulating effect.

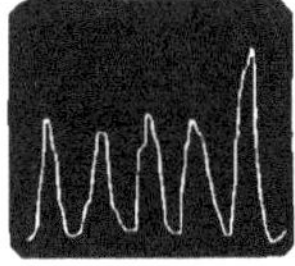

Fig. 54. Influence of Musk.

Joy and anger increase the power, sorrow and fright decrease it. An entertaining novel is a will-stimulant; a prosy text-book actually weakens us.

CHAPTER VII.

ATTENTION.

AT-TENtion! Why do you suppose such a command is necessary to a company of soldiers or a class of boys? Would they not do as well without attention? Of course not; an inattentive or unexpectant company of soldiers will start to march in utter disorder or will ground arms with a running fusilade of bangs instead of a single thump.

What is attention?

What is this difference between attention and inattention, between expectation and surprise? How can we turn inattention into attention?

In the first place, What is attention? It is a very sad fact, but I cannot tell what it is. The innumerable psychologies attempt to define it, but when they have defined it, you are sure to know just as much about it as before.

Illustration.

When you first move into a new neighborhood, you notice every house, every tree, almost every stone, as you pass to and fro. As you grow accustomed to the surroundings, you gradually cease to notice them. Finally you pay so little heed to them that you are unable at the end of a walk to tell what you have just seen by the way. This fact is expressed by saying that at first you attended to what you saw and afterwards did not.

Another illustration.

I can illustrate this process of attention in another way. You are now reading the sentences on this page; you are giving full attention to what I say. But at the same time you are receiving touch impressions from the book in your hand and from the clothes you wear ; you hear the wagons on the street or the howling of the wind and the rustling of the trees ; you smell the roses that your hostess has placed on the table. Now that I have mentioned them you notice them—or pay attention to them. When you were attending to what you were reading they were only dimly present.

Focus and field of attention. I will suppose that you are attending to what you are reading ; all those sounds, touches, smells, etc., are only dimly in the *field* of your experience while these words are in the *focus* (or burning-point) of experience.

Illustration from the camera. Probably you can gain a good idea of the difference between the focus and the field of present experience by taking an analogy from the art of photography. Ask your friend the amateur photographer to bring around his camera. He sets it up and lets you look at the picture on the ground glass. The glass is adjusted so that the picture of a person in the middle of the room is sharply seen ; all the other objects are somewhat blurred, depending on their distance from him. Change the position of the glass by a trifle. The person becomes blurred and some other object becomes sharp. Thus for each position of the glass there is an object, or a group

Fig. 55. Focus and Field of Attention.

of objects, distinctly seen while all other objects are blurred. To make one of the blurred objects distinct, the position of the glass must be changed and the formerly distinct object becomes blurred.

In like manner, we fully attend to one object or group of objects at a time; all others are only dimly noticed. As we turn our attention from one object to another what was formerly distinct becomes dim.

Instability of the focus of attention.

The illustration with the camera is not quite complete. You can keep the objects quiet in the room but you cannot keep your thoughts still. The mental condition would be more nearly expressed by pointing the camera down a busy street. You focus first on one thing, then on another. The things in focus pass out of it, others come in. Only by special effort can you keep a moving person or wagon in focus for more than a moment.

Instead of talking all around attention, as psychologists have been doing for two thousand years, let us ask a few practical questions. The possibility of answering some of them arises from the fact that we can now experiment on attention. The impossibility of answering the others is due to the lack of psychological laboratories and trained psychologists to make more experiments.

Extent of attention.

In the first place, How many objects can be attended to at a time? Objects can, of course, be more or less complicated. A house, for example, is a single object if we do not look into the details; it is a multitude if we count the windows, doors, roof, chimneys, etc., as separate objects. By the word "object," then, we will understand any thing or group of things regarded as a single thing. Thus, the natural tendency would be to regard the letters M X R V as four objects, four letters, whereas MORE would be regarded as one object, a word, unless we stop to consider the letters separately.

Experiments.

Experiments are made by exposing pictures, letters, words, etc., to view for a brief time. One way of doing this is to prepare slides for the projection-lantern and throw the view on the screen for an instant.

A more convenient way is to fix the pictures or letters on cards or to prepare a table on which actual objects are placed. A photographic camera with a quick shutter is aimed at them. The person experimented upon is so placed that he cannot see the objects but can see the ground glass.

If you cannot use a lens with shutter, the experiment can be tried roughly in the manner explained on page 22 ; the time of sight of the card must, however, be less than one tenth of a second.

Experiments of this kind show that four, and sometimes even five, disconnected letters, numerals, colors, etc., can be grasped at the same time. When the objects are so arranged that they enter into combinations that make complex objects, many more can be grasped. Thus, two words of two syllables, each word containing six letters, can be grasped as readily as four single letters.

Extent of attention in children.

This ability to grasp and remember complicated objects increases with age. Children seem to grasp only the details separately and to be unable to gain a general view with the parts in proper subordination. In drawing a horse unskilful persons will begin with the head, proceed with the back, then the rear legs, etc., thinking of only one thing at a time ; the result is generally that the various parts are out of all proportion. The details are often so isolated in the child's mind that he will draw parts entirely separated from one another.

This is the case with the child that drew an oblong and a square separately to stand for the two sides of a box seen in perspective.

Methods of forcing attention.

Let us consider first the methods of forcing attention to an object, or, as is frequently said, of forcing the object into attention.

The law of bigness.

The first law I shall state is : *Bigness regulates the force of attention.* Young children are attracted to objects by their bigness.

Used in advertising.

Advertisers make it a business to study the laws of attention. American advertisers in the past and also largely in the present rely chiefly on the law of bigness. They know that one large advertisement is worth a multitude of small ones. A certain New York life insurance company puts up the biggest building ; *The New York World* builds the highest tower. Churches frequently vie in building, not the most beautiful, but the largest house of worship.

Curiously enough, the rage for notice even invades the solemnity of death. Joseph Frank was not content to have his ashes rest in peace on the shore of the Lake of Como ; he must erect a pyramid to attract the attention of travelers. Richard Smith lately bequeathed $500,000 for a big memorial arch in Philadelphia.

The cost of attracting attention by bigness.

Bigness, however, costs. The art of successfully applying this law of bigness lies in finding the point at which any increase or any decrease in size lessens the profit. For example, let us suppose that we have manufactured a new kind of cloth. As long as nobody pays any attention to the matter, nobody buys. We determine to spend $1,000 in advertising by a brief notice for a large number of times in the regular type of the paper. Among the numerous other advertisements ours attracts no attention ; the money is wasted. We try again, putting in half as many advertisements but making each twice as large. We get a better return. By continually increasing the size at the expense of the number of repetitions, we get steadily better returns till the bigness of the advertisement is sufficiently striking to render any increase unnecessary. Any further increase does no good by reason of its size, but does injury by decreasing the number of repetitions. The skilful advertiser will stop just at the maximum point.

Lack of scientific investigation of this law.

It is a curious fact that the keenness of business men often leads them unconsciously to anticipate the discoveries of science. The law by which the intensity of attention is related to bigness has never even been proposed in the psychological laboratory, yet the successful advertisers have learned it by practical experience. The law I have here explained in popular form and the laws I am about to mention should and must be made the subject of the most careful, most accurate, and most extended investigation in the psychological laboratories. Every detail, every application, must be sought for. There is no more vital question in all mental science than this one of attention. The whole subject of interest, about which we are accustomed to hear so many Herbartian platitudes, is merely one phase of it. The scientific psychologists are deeply to blame—as I am included in the reproach I can speak freely—for not having by experiment and measurement rescued this chapter of all chapters from the clutch of the old psychology.

Law of intensity.

The second law of attention which I venture to propose is the *law of intensity or brightness*, according to which *the intensity of a sensation influences the amount of attention paid to it.* Here also we have no experimental results ; we must, for our examples, rely on the art of psychology rather than on the science.

Examples.

The shopkeeper well knows the effect of a gilded sign. The druggist's bright light forces you to notice him. The headlight on the trolley-car serves another purpose in addition to lighting the track. The Madison Square Garden in New York advertises itself by its bright lights.

The clanging gong, the excruciating fish-horn, the rooster's crow, and the college yell are all for the purpose of attracting attention. **Full black letters** for paragraph headings or advertisements are more effective than ordinary type or outline letters.

Cleanliness is not the only reason why a man-of-war is kept in a high degree of polish. The furnishings could be just as clean if painted with black asphalt, but the effect on the officers and men would be quite different. It is impossible to get full attention to duty and discipline in a dingy vessel.

Application to teaching.

This same principle is applied in instruction. An old or rusty piece of apparatus cannot command the same attention from the students as a brightly varnished and nickel-plated one.

Students in a chemical laboratory do not pay nearly as good attention to their manipulations if they work over scorched, stained tables and black sinks. The director of one laboratory in Belgium covers his tables with fine, white lava-tops. The expense is at first great, but the increased attention more than repays the cost. Experience has shown that the students working at those tables keep their glassware cleaner and do their chemical work with more care than those who work at the ordinary wooden tables.

We noticed the use of bigness for memorial purposes ;

the use of brightness is also common. The brilliancy of stained glass windows attracts at the present day as much attention at a moderate expense as could be attracted by a costly, beautiful statue or tablet.

Law of feeling.

The third law I shall call the *law of feeling;* it can be stated in this way: *The degree of attention paid to an object depends on the intensity of the feeling aroused.* The feeling may be either of liking or disliking.

Painful sensations arouse a strong dislike. "The burned child dreads the fire"; it is equally true that a burned child watches the stove. The very name of croup strikes terror into the mother and the slightest hoarseness arouses her attention.

Few feelings are so intensely pleasurable as those of the young mother. Watch the *tension*—the *at*tention—to every movement of the child.

Beauty and bigness.

In former days beautiful objects were accompanied by intensely pleasurable feelings. When Giotto wished to give Florence a remarkable tower, he made it of wondrous beauty. When the Parisians wished a striking tower for their exposition, they got M. Eiffel to make it the tallest one.

To celebrate the victory of his chorus in the theater of Bacchus, Lysikrates erected in Athens his famous choragic monument. Exquisitely wrought, graceful in its proportions, rich in decoration, perfect in its material, it is the wonder and admiration of the world. True, it is only thirty-four feet tall, and to-day in competition with the Ferris wheel would not attract the slightest attention—unless it could be used as the ticket-office.

Various feelings employed.

In fact, our crude western civilization, our puritanical love of the ugly, and our color-blind Quakerism have deprived us of any feeling for beautiful objects. If an appeal is to be made for attention through feeling, it must be done in some other way. The other way is generally by use of the comic, the grotesque, or the hideous ; for example, the *soi-disant* jokes that are interspersed all through our newspapers, the cartoons of *Puck*, and the colored supplements of the New York Sunday papers.

Personal pride.

Personal pride is accompanied by strong feeling which brings attention to anything necessary for its proper maintenance. Vanity is closely connected with attention to dress. The personal pride may extend to our club, our town—nay, even to our country; for not all patriots are scoundrels, some are merely vain.

It was this same *esprit de corps* that Bonaparte knew how to arouse. Bismarck and Moltke won the Franco-Prussian War by the attention of every soldier to his duty.

Artificial interest.

The culminating point in education is the power to attend to things that are in themselves indifferent by arousing an artificial feeling of interest. There is hardly anything of less intrinsic interest to the student than analytical mathematics, such as algebra; the treatment is purposely deprived of every concrete relation. Yet we know that the power of attending to such a subject can be cultivated. Indeed, I have heard that there are some mathematicians who even take an interest in that science.

Law of expectation.

The fourth law of attention which I shall propose is the *law of expectation*—I had almost said, of curiosity.

A step at the front door arouses a memory of a bell-ring; the ear is prepared to hear it. Whether the matter concerns us or not, this condition of expectation forces our attention.

The peacock who lived next door to De Quincey almost maddened him by the expectation of the coming scream. The actual scream was a relief; thereafter the attention became steadily more and more intense till the moment of the next scream. The law of expectation is used in a masterly way in Dickens's "Mutual Friend." It is a characteristic of successful newspaper writing that the opening paragraph shall arouse expectation, and therefore attention. The same principle underlies the art of writing headlines.

Expectant attention.

Curiosity is expectation where the mental picture is very indefinite. We all know the story of P. T. Barnum and the brick. We can also understand why *The New York Herald* put large glass windows in its publication building.

Scientific curiosity.

Scientific men are famed for strict and ardent attention to their investigations. The fascination of research and discovery lies in the vague expectation of something new. The essence of all science is curiosity —the same every-day, good old homely curiosity that impels Farmer B——'s wife to learn just how many eggs are laid by her neighbors' hens, that makes Robbie pull apart his tin locomotive to see how it works, or that induces kitty to stick her paw down a knot-hole in the floor. The next time a scientific man quotes that scandalous — but true — proverb about curiosity and woman, let my fair readers ask him, if he is a zoölogist, why he pries into the housekeeping habits of the squirrel (Farmer B——'s wife); if he is a botanist, why he pulls your prettiest flower to pieces (Robbie); if he is an archæologist, why his friends so attentively poke themselves into the pyramids and tombs of Egypt (kitty and the knot-hole).

Unsatisfied curiosity.

Unsatisfied curiosity arouses still more attention. Many papers still maintain puzzle columns, well knowing that unsatisfied curiosity is a more intense form of unsatisfied expectation. Possibly the strange, complicated designs on our magazine covers are meant to be puzzles that can never be solved. The reason of the great attention paid to Stockton's "The Lady or the Tiger" is to be found here.

It is a principle of serial stories that each installment shall end with an unsatisfied expectation. This contributes more than the merit of the story to arousing the attention of the reader, who, because he keeps thinking of what may happen, is forced to buy the next number of the periodical in order to be relieved of the tension.

Law of change.

The fifth law of attention is the *law of change*, or the law of unexpectedness; *the degree of attention depends upon the amount and on the rapidity of the change.*

Things indifferent and even things unpleasing may leave their impress by the severity of the shock they give. A flash of lightning or a low door-lintel commands notice. There is an old saying that wonder is the beginning of philosophy. Various things may be meant by wonder, but one thing is the shock of mere surprise or astonishment; at any rate an impression is made.

In our reading we expect only straight lines. The advertiser arouses attention by tipping them slantwise. The advertiser makes frequent use of this law combined with the law of curiosity by putting in his notice upside down.

Effect of attention on mental quickness and regularity.

A prominent effect of attention is to shorten reaction-time and thought-time and make them more regular. The commands of a military captain are really signals for reaction. The first part of a military command is arranged to serve as a warning signal to insure good attention; "Shoulder—ARMS!" "Right—FACE!" The acts of the men are simple reactions. They are not associations; therein lies the reason why a command is not given as a single expression. If the command were "Forward-march," delivered as one expression, the soldier would be obliged to discriminate, associate, and choose among twenty or thirty possible commands. We have already seen that these processes not only take a very long time but are very irregular; moreover, no warning would have been given. The company could not possibly step forward as one man. Whereas the command "Forward—MARCH" causes all the mental processes except simple reaction to be performed beforehand; every man in the company has but one thing to do, his attention has been properly prepared by the warning and the whole company should start together.

Attention and pride.

I venture to suggest that the splendid drilling of the Seventh Regiment, N. G. S. N. Y., is due to the intense attention paid to the commands. Although the men are under drill only once a week, they compete with and often surpass the regular troops, who drill several times a day. I know from personal experience that the regimental pride was something stupendous and that while under drill the mind was tuned to a high pitch of excitement. Every thought was on the captain; the eye and the ear were strained to catch the next command; every muscle was waiting its orders from the will. In fact, it often seemed as though the muscles obeyed the captain's orders without waiting for the man's own.

The use of the preparatory signal for the purpose of fixing attention is familiar in the sailors' cry, "Yo—Ho!" whereby they can pull together.

A notable effect of attention to one idea is the lack of

attention to other ideas. Henry Clay was obliged to speak on one occasion when in very delicate health. He asked a friend who sat beside him to stop him after twenty minutes. When the time had passed, the friend pulled Clay's coat, but he continued to speak. His friend pinched him several times and finally ran a pin into his leg. Clay paid no attention. He spoke for more than two hours and then, sinking exhausted, he upbraided his friend for not giving him a signal to stop at the proper time. The signals had been given but his mind was so intensely attentive to his discussion that everything else was neglected. It is a well-known fact that we can forget griefs, pains, even the toothache, when reading a fascinating book or watching a forcible drama.

Concentration of thought.

Excessive cases of this effect of attention are seen in the men of one interest and the men of one idea. We have men who will listen to nothing but discussions of Shakespeare, others whose sole idea lies in pork.

Men of one idea.

Going still further we find abnormal cases: arithmomania, where the patient is continually asking why the houses are so large, why the trees are so tall, or where he is continually counting the number of paving-stones in the street or the number of rivers in a country; metaphysical mania, where the patient cannot hear a word like "good," "beautiful," "being," etc., without irresistibly speculating on the problems of ethics, æsthetics, and metaphysics. These and similar cases are included under the term of "fixed ideas." The acute stage of excessive attention is found in ecstasy.

Diseases of attention.

The methods of rapidly fatiguing attention have lately been brought into notice by hypnotic exhibitions. Preparatory to influencing a person by suggestion he is often

Fatiguing attention.

reduced to a half-dazed condition by steady gazing at a bright object, by repeated bright flashes, by monotonous noises, by regular strokes of the hand, etc. This process consists essentially of a fatigue of attention. It is generally called "hypnotizing." The name seems justified, as the resulting condition resembles the somnambulic form of sleep where the patient is half awake.

Hypnotizing.

Fig. 56. Fatiguing Attention Preparatory to Hypnotism.

Fatiguing attention to produce sleep.

The means employed are close copies of well-known methods of avoiding sleeplessness. Steady gazing at the ceiling, the tick of a watch under the pillow, the hum of a dynamo on shipboard, the roar of the falls and the grinding of the mill, the stroking of the invalid's brow—these have banished many an hour of hopeless tossing.

CHAPTER VIII.

TOUCH.

HERE is a row of ten little disks, 3^{mm} in diameter, cut from elder-pith. Each is suspended by fine cocoon-fiber from a little handle. For portability the handles are stuck in holes in a support, Fig. 57. Now place your hand comfortably on the table and close your eyes. Tell me when and where you feel anything touch your hand. Without letting you know what I am doing I take the handle with the lightest weight and let the weight softly down till it rests on your hand (Fig. 58). You do not know that I have done so, and you feel nothing. Then I try the next heavier, and so on, till you feel the pressure. The little disks are graded in weight, thus 1^{mg}, 2^{mg}, etc., up to 10^{mg}.

An experiment on touch.

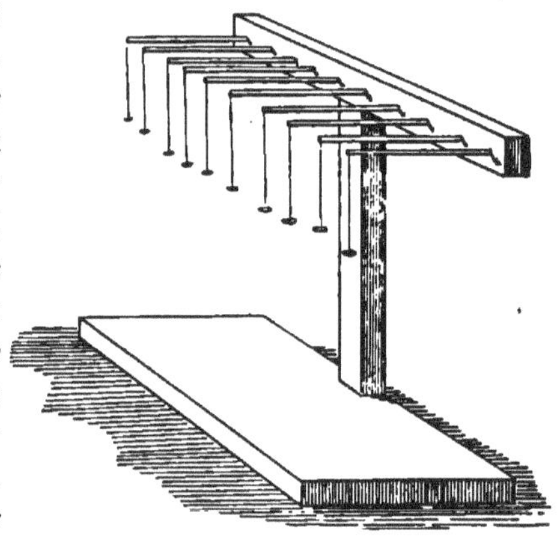

Fig. 57. Touch-weights for Finding the Threshold.

Now, if the fourth weight was the first you felt, then 4^{mg} was the least noticeable weight, or the weight just on

Threshold of touch, or pressure.

the threshold of intensity. This fact of the threshold is one that we shall meet everywhere in the study of mind.

The threshold of sensation for the sense of pressure in an average subject was 2^{mg} for forehead, temples, and back of forearm and hand; 3^{mg} for inner side of forearm; 5^{mg} for nose, hip, chin, and abdomen; 5^{mg} to 15^{mg} on inner surface of fingers; and $1{,}000^{mg}$ on heel and nails.

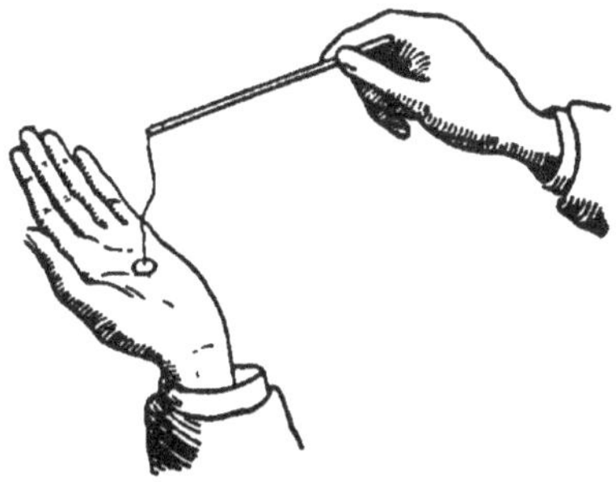

Fig. 58. Finding the Threshold for the Palm of the Hand.

A second experiment.

Some idea of the delicacy for distinguishing differences in pressure can be obtained by laying a hair on a plate of glass and putting over it ten to fifteen sheets of writing paper. The position of the hair can easily be felt by passing the finger back and forth over the surface.

Touching with movement gives much more delicate judgments than mere contact. A book-cover feels much rougher when the finger is moved over it than when it is merely touched.

Tickle.

Something very peculiar occurs when a light pressure is varied rapidly in intensity. If the tip of a tuning-fork in motion be slowly touched to the skin, it "tickles."

Use of the tuning-fork.

A tuning-fork when in motion shakes (or vibrates). A tuning-fork can be made to record just what it does when it shakes. Glue a hair to the end of the fork. Smoke a piece of window-glass in a candle-flame for a moment, moving it about to keep it from cracking. It will soon be covered with a layer of smoke. Holding the fork by the stem, set it shaking by striking it smartly across the knee or edge of a flatiron wrapped in several layers of cloth. With a quick movement, draw the fork so that the hair traces a line in the smoke. A curve will be drawn like that in Fig. 4.

Now, if the fork be held with the end touching the skin, as in Fig. 59, it is plain that the hair must be producing a fluctuating pressure. The result is an unbearable "tickle." This peculiar form of pressure can be called a wavy pressure. A light, wavy pressure, then, produces a tickle.

Wavy pressure.

The tickling pressure need not be a true wavy pres-

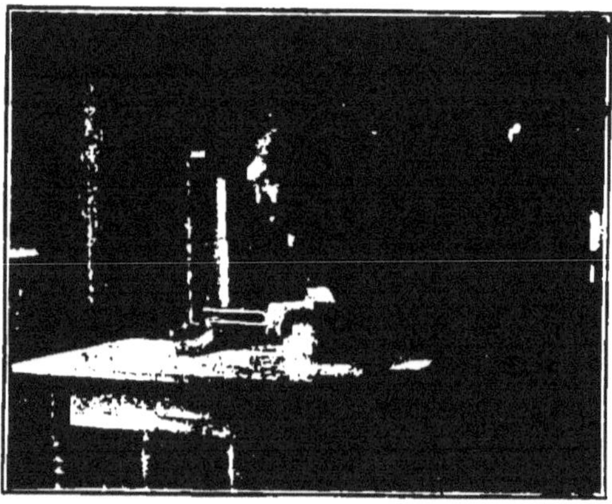

Fig. 59. An Experiment in Tickling.

sure ; that is, it need not be perfectly regular. If any object, such as a feather or the finger, be held lightly against the face, a tickle is felt, due to the trembling of the hand.

Tickling pressure need not be regular;

The tickling thing need not stay at one spot, but may be moved along continuously. A feather drawn over the temples makes a strong tickle. A fly walking over

or at the same spot.

the skin produces an unbearable tickle in exactly the same way. Stories of the Thirty Years' War relate how the soldier-robbers forced the peasant to reveal his treasure by subjecting him to unbearable tickle.

Change of pressure.

When a pressure is already felt, it can be made stronger or weaker to a certain degree before the change is perceived.

Experiment.

The experiment can be made with a pair of beam-balances. The hand, supported by a block or cushion,

Fig. 60. Finding the Least Noticeable Change in Pressure.

is placed under the scale-pan so that when the scale is at rest, the pan-holder just touches the skin (Fig. 60). To avoid the coldness of the pan, a piece of cork or leather is placed between the hand and the metal.

The subject of experiment closes his eyes. A weight

is placed in the pan above the hand. A sensation of pressure is felt. Sand is quietly poured into the same pan until the subject feels the pressure to be increasing. By putting weights in the other pan the amount of increase can be measured. Now start with the same weight as before and pour sand into the opposite pan until the subject feels the pressure to be lighter. The amount of sand that has been added represents the least noticeable *change*, or the threshold of change, in the pressure. Thus, if the weight at the start was 50^g and the amount of sand added was 35^g, the least noticeable change was 35^g, or $\frac{35}{50}$ of the original pressure.

Least noticeable change.

Influence of rate.

Several facts will be noticed by those who perform this experiment. In the first place, the least noticeable change depends on the rate at which the change is made. Several funnels should be used, with the ends of different sizes. When one of these is filled with sand, the rate at which the sand flows out depends on the size of the opening; some funnels will allow the sand to flow rapidly, others slowly. When the same experiment is repeated with different rates of flow, it will be found that the slower the flow the greater the least noticeable change. With a very slow flow the weight can often be increased two or three times over before the change is noticed.

An extremely slow rate.

No one has ever tried to see if a great pressure can be applied to the human skin without its being noticed, provided the rate be extremely slow. A frog with the spinal cord cut off from the brain is quite sensitive to a touch; yet when a pressure is applied by screwing a rod down at the rate of 0.03^{mm} in one minute his foot can be crushed in $5\frac{1}{4}$ hours without a sign that the pressure was felt.

The next point to be remarked is that the least notice-

Influence of the starting-point.

able change depends on the weight from which the pressure is started. Roughly speaking, if for a weight of 50^g the least noticeable change, at a certain rate, is 30^g, or 60 per cent, then the least noticeable change, at the same rate, for 25^g will be 15^g, or 60 per cent, not 30^g.

General law.

These two classes of facts can be summed up in one general law: The threshold of change increases inversely as the rate of change but proportionately as the starting pressure.

Strangely enough, although change and rate of change enter into nearly every experience of life, almost nothing has been done in the experimental study of the subject. Several years ago I called attention to the importance of this factor of the rate of change. Here, for the first time, I have taken the liberty of proposing the law of the threshold of change. It is based on various observations I have made at different times; nevertheless, no extended investigations on the subject have been made, and until these are done the law cannot be regarded as definitely established.

Least noticeable difference.

The least noticeable difference is quite another matter from the least noticeable change. The usual method of experiment employs a series of weights successively growing slightly heavier or lighter from the standard.

Experiment.

Suppose we start with a weight of 20^g as a standard, and have a set of weights increasing or decreasing successively by steps of 1^g. The standard is first applied, say, to the palm of the hand—the hand being at rest on a cushion. It is then removed and, after about two seconds, the 21^g weight is applied for an instant. The subject tells whether he feels it lighter, heavier, or the same. After a short time the standard is again used; then the 22^g weight is applied. This is continued with 23^g, 24^g, etc., till the subject has several times in succession felt the weights to be heavier. The first weight of the unbroken succession of heavier weights gives the least noticeable difference. For example, suppose a set of experiments to give the following results: 21 equal, 22 heavier, 23 lighter, 24 equal, 25 heavier, 26 heavier, 27 heavier, 28 heavier. Then the threshold would be at 5^g, all differences less than 25 — 20 being uncertain.

Various standards. In a similar manner the threshold of difference can be found with successively lighter weights. For a general threshold the average of the two can be taken. For example, if the threshold for 20^g toward lightness is 4^g and the threshold toward heaviness is 5^g, the average threshold is $4\frac{1}{2}^g$. When different weights are used as standards, it quickly becomes apparent that the threshold of difference does not remain at the same number of grams. For a standard of 200^g the difference of 5^g will not be felt at all. The threshold will be more nearly 20^g.

Results. The results of such a series of experiments are given in the following table:

S	1	2	5	10	20	50	100	200	500	1,000	2,000	4,000
D	0.2	0.3	0.6	0.9	1.5	2.8	6.4	10.8	25	57	80	100
$\frac{D}{S}$	$\frac{1}{5}$	$\frac{1}{7}$	$\frac{1}{8}$	$\frac{1}{11}$	$\frac{1}{13}$	$\frac{1}{18}$	$\frac{1}{16}$	$\frac{1}{19}$	$\frac{1}{20}$	$\frac{1}{18}$	$\frac{1}{25}$	$\frac{1}{40}$

The figures in S give the different standards; those in D give the least noticeable differences; those in $\frac{D}{S}$ tell the relation of the least noticeable difference to the standard. Thus, for a standard of 1^g the least noticeable difference is 0.2^g, or $\frac{1}{5}$ = 20 per cent. For $1,000^g$ it is 57^g, or $\frac{1}{18}$ = 5.7 per cent.

Weber's law. It is evident that the least noticeable difference does not remain the same but increases as the standard increases. The famous law of Weber would say that the least noticeable difference increases in the same ratio as the standard; in other words, that the least noticeable difference is always a certain fraction of the standard. This is not true for pressure, as is seen by the line of fractions for $\frac{D}{S}$; according to Weber's law they should all be the same.

This law of proportionality of differences is recognized in many tax laws. For example, the income tax demands that each person shall pay an amount in direct proportion to his income. The Mosaic tithe demanded a tenth. This is presumably all in the belief that a man with $100 feels a payment of $10 as much as a man with $100,000 feels one of $10,000.

Law is too simple.

In saying that like differences are not differences of the same amount, but are differences depending on the amount from which you reckon, the law is unquestionably true. But the relation of proportionality is much too simple to meet the facts.

It is a curious and interesting fact that much finer differences can be detected when the two weights are applied one to each hand at the same time.

Threshold of space.

Let us find the threshold of space for the skin. An ordinary pair of drawing-dividers can be used, but accurate work requires a better apparatus. The compass in Fig. 62 consists of a horizontal bar on which slide the two points. These points are held on springs so that the experiment can be made at a constant pressure.

Fig. 61. Simple Æsthesiometer.

Experiment.

Place the two points at 1^{mm} apart. Take the æsthesiometer by the handle and gently press the points against the forehead of some one who has his eyes closed and who has not seen the adjustment of the points. He is to say whether he feels two points or one. At this distance he will feel only one. Adjust the points to 2^{mm} and try again. Proceed in this way till he feels the two points distinctly. Now start with a somewhat greater distance and proceed backward till only one point is felt. The average of the two results is the threshold of skin-space at the particular pressure for the particular place of the particular person experimented upon.

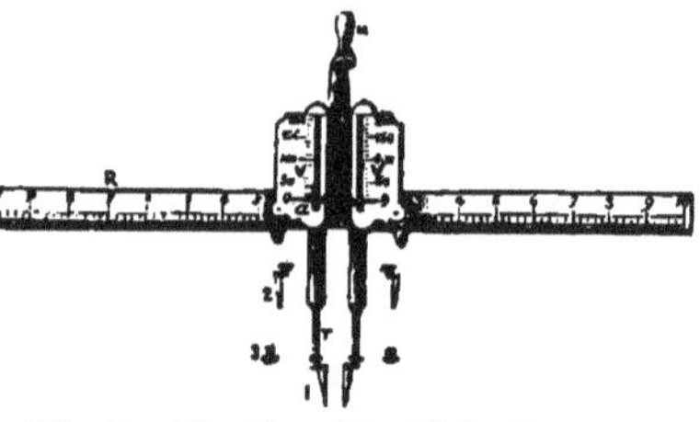

Fig. 62. The Complete Æsthesiometer.

Results.

Here is a specimen table of results from Weber :

Tongue	1^{mm}
Inner side of first finger-joint	2^{mm}
Lips (red portion)	5^{mm}
Inner side of second finger-joint	7^{mm}
Lips (skin)	9^{mm}
Cheek, big toe	11^{mm}
Forehead	23^{mm}
Back of hand	31^{mm}
Leg	40^{mm}
Neck	54^{mm}
Middle of back, upper arm, thigh	68^{mm}

Education of the skin.

It is a remarkable fact that the skin can be educated by practice so that the threshold is much reduced. This can be measured directly by Weber's compass ; any one with a pair of dividers can try the experiment on himself.

Threshold of space among the blind.

The blind, who pay constant attention to their finger-tips, have very small thresholds. Curiously enough, their thresholds are also smaller on the back and on other places which they do not use more than other people. The superiority of the blind in this respect would seem to be due to increased attention to the skin. A further evidence of this explanation is the fact that education of one part of the body brings a special decrease of the threshold for the neighboring parts and for the same portion of the opposite side of the body. The experiment can be performed in this way: First find the threshold for the front of the wrist of the left arm, trying it five times; then find it for the same place on the right arm, trying for ten minutes; then on trying again on the left arm the threshold will be found to be less.

Cross-education.

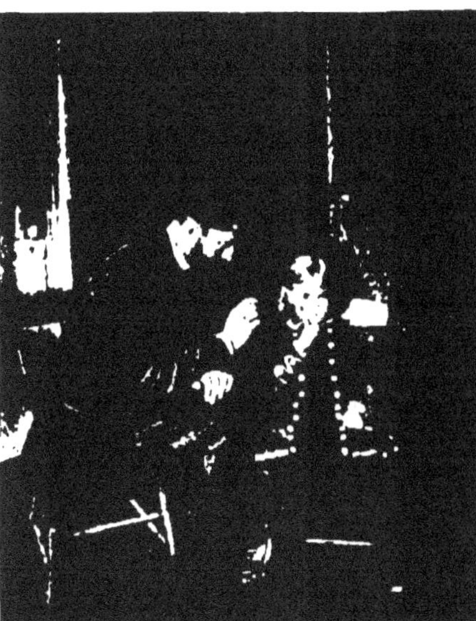

Fig. 63. Testing a Child's Idea of Skin-space.

Location of points on the skin.

The fineness of distinction for space on the skin can be shown in a simpler but less accurate way. The person experimented upon closes his eyes. Some one touches him with a pencil point and he moves another pencil to the point where he was touched. The error is measured.

There is an interesting application of this experiment

Experiment with a child.

which I will propose to mothers. Young children cannot be made to understand either of the last two experiments. An intelligent and patient mother, however, can teach her child, even before it can talk, to put its finger to the spot on which it is touched. Dip the end of the child's finger into something black, *e. g.*, soot, pencil-filings, powdered graphite, or blacking. Touch the child with a pencil and let him point to the spot. Measure the distance between the pencil mark and the finger mark.

Perplexing the skin's idea of location.

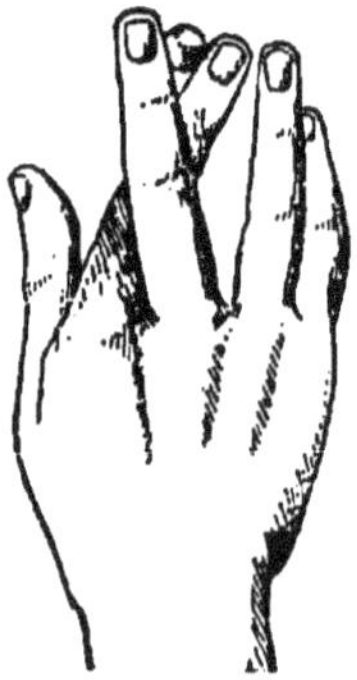

Our experience has taught us that the various portions of the skin stand in certain space-relations. Thus we know that something touching the middle finger is further from the thumb than something touching the index finger. When the fingers are out of their places we are irresistibly driven to judge as if they were in proper order. This is illustrated by what is known as Aristotle's experiment. The middle finger is crossed over the index finger in such a way as to bring the tip of the middle finger on the thumb-side of the other. A pea or other small object, when inserted between the two, will appear as two objects. It is difficult to re-learn the arrangement of the skin in space. We thus see why a person whose nose has been re-formed by a piece of skin from the forehead, for a while feels all contact on the nose as if it were contact on the forehead.

Fig. 64. Aristotle's Illusion.

The finger illusion.

The lip illusion.

A similar illusion is produced by placing a pencil between the lips and moving the under-lip to one side. There are apparently two pencils.

The distortion of space under unusual conditions is familiar to persons in the dentist's chair.

Interrupted space.

The distance between two points on the skin seems

Fig. 65. Lip Illusion: 1. The Reality; 2. The Feeling.

greater when the skin between these points is also touched. If four pins are pounded in a straight line into a stick at one fourth of an inch apart, the distance between the end pins will appear greater than that between two separate pins three fourths of an inch apart.

Fig. 66. Space as Estimated by a Tooth under Treatment.

Smoothness and roughness.

The distances apart of the various points that we feel are what we know under the names of smoothness and roughness. A billiard-ball is "smooth," that is, our sensations of touch are evenly distributed. Carpet is "rough," that is, it produces uneven sensations. Sandpaper is peculiarly "rough," because very intense and limited sensations from the sharp sand are mingled with smoother ones and gaps. Velvet, when felt backward, has a peculiar rough smoothness, because the separate points of the individual hairs produce separate sensations, yet they are so near together as to resemble smoothness. Short-nap plush has a similar feeling. The smoothness of baby's cheeks can be contrasted with the skin of the inhabitants of Brobdingnag.

Ideas of form.

If a little cardboard triangle, circle, or square be laid on the hand and pressed down by the point of a knife or a pencil in the center, we get a combination of pressures from every point on the surface. Certain combinations are said to belong to triangular objects, certain others to circular objects. The pressures thus not only represent a quality of surface but also of form.

As the judgment of distances is limited to distances larger than the threshold, very small cardboard forms all appear as points.

CHAPTER IX.

HOT AND COLD.

The old idea of hot and cold.

IN THE old days it was supposed that heat and cold were two different things; even to-day the uneducated person cannot grasp the idea that coldness is simply the absence of heat, that a piece of ice is cold simply because it is not hot. But the modern development of physics has shown that heat consists of motion among the little molecules of which all bodies are supposed to be composed, and that as this motion becomes less the bodies are said to be cold. Thus, a glass of warm water differs from a glass of cold water simply in the fact that the molecules of the water in the former are moving rapidly, while in the latter they are comparatively quiet.

The physical idea.

The psychological idea.

Strange as it may seem, it was discovered a few years ago that the ordinary common sense of every-day people was right. Not that the science of physics was wrong, but that the conclusion drawn was incorrect. Hotness and coldness are two entirely different things from our point of view. A glass of water is warm because it gives us a feeling, or sensation, of warmth; another glass is cool because it gives us an entirely different sensation of coldness. The complete distinction of our feelings of hotness and coldness from the physical condition of the molecules of the object touched is emphasized by an experiment in which the same object feels both hot and cold at the same time.

Hot spots and cold spots.

Our sensations of hot and cold come from little spots called hot spots and cold spots. To find the cold spots a pointed rod, *e. g.*, a lathe center, a pointed nail, or even a lead pencil, is cooled and then moved slowly and lightly over the skin. At certain points distinct sensations of cold will flash out, while elsewhere nothing but contact or vague coldness is felt. These points are the cold spots; a specimen arrangement of them is shown in Fig. 67.

C

Fig. 67. A Cold-spot Map.

To find the hot spots the metal point is heated and applied in a similar manner. The hot spots are everywhere different from the cold spots. A specimen case is shown in Fig. 68.

Making a cast of the hand.

At the art store get a few pounds of plaster for casts (the finely ground plaster, not the ordinary plaster of Paris). Mix it with water in a bowl. Pour out a portion into a tin pie-plate. Now press the hand (palm or back) down upon it, being careful to touch the plaster at every point. When the plaster has hardened sufficiently to permit the removal of the hand without sticking, carefully raise it. A perfect cast of the hand is obtained with every line expressed.

Fig. 68. A Hot-spot Map.

Finding the spots.

Now prepare yourself with a glass of ice water, a glass of hot water, some red and some blue ink, a pointed metal pencil (or a sharp lead pencil), and a couple of toothpicks. Cool the pencil in the ice water. Dry it and pass it over the skin. Whenever a cold spot flashes out, mark its position in blue ink with a toothpick on the cast. The fine creases in the skin will enable you to locate it exactly. Repeat this a few times, till you are satisfied that you have a map of all the cold spots. Warm the pencil in the hot water and find the hot spots in the same way. Mark them on the cast in red ink.

Permanent record. When you have finished you will have a complete geography of your temperature spots on a relief map. Separate the cast from the pie-plate ; make a plush frame for it, and hang it up in your art gallery. Those of you

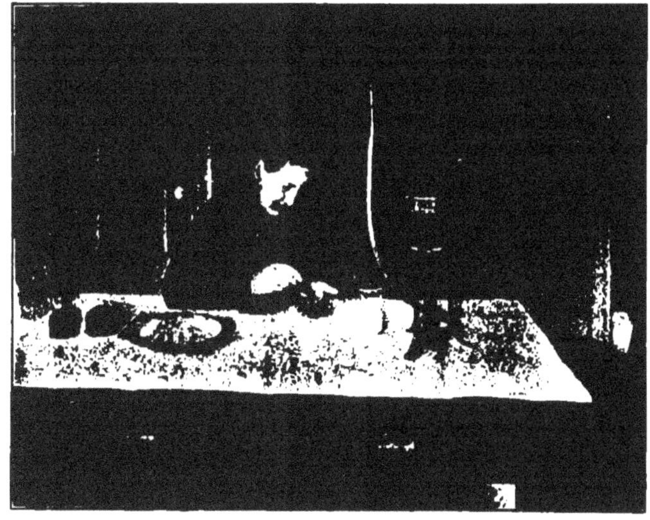

Fig. 69. Finding the Hot and Cold Spots.

who are willing to do a trifle more work can use the impressed cast as a mold from which to get a hand in actual relief. Very few of us can afford a gallery of statues of ourselves to be transmitted as remembrances to our descendants. Why should not such collections of hands, with their hot and cold spots, be found in future centuries in the ancestral galleries of our posterity? Mental peculiarities are of as much interest as oddities of dress ; indeed, to our descendants they are of far greater interest and importance. Any one who is willing to give a little more time to the matter might find out the threshold of touch (page 103) for various places on the hand and mark the number of milligrams on the cast.

Automatic method of locating the spots.

In very accurate work we are troubled by the impossibility of keeping the metal pencil at anywhere near the same temperature and by the uncertainty in marking the spots. To overcome these difficulties I have invented an instrument for mapping the hot and cold spots on the skin. It consists of a pointed copper box whose sides are protected by felt. Through this box there runs a steady stream of water from a reservoir. The water in this reservoir is kept at just the same degree of heat by means of an automatic regulator of the flame. A thermometer in the copper box tells what the temperature of the point is. By adding cold water from another reservoir, we can use any temperature we desire. The little copper box is made to travel over the hand ; as it does so a pencil travels just above a piece of paper. Whenever a spot is felt, the person presses a telegraph key with the other hand ; by means of a magnet this causes the pencil to strike downward and make a dot on the paper. By these means an accurate map is automatically made.

Always hot and cold from the same spots.

The hot spots are ordinarily not sensitive to coldness or the cold spots to heat. Yet a very hot point applied to a cold spot so as not to reach hot spots also will feel cold ; of course, to a hot spot it is intensely hot. It is noteworthy that when the hand is applied to a very hot or a very cold object there is often doubt for a few moments whether it is hot or cold.

The temperature spots answer to tapping by sensations

of hot or cold. For example, choose a sensitive cold spot and let some one tap it with a fine wooden point ; it will feel cold. Thrust a needle into it ; it will feel no pain.

Law of change for hot and cold.

In studying the subject of touch we had occasion to notice a certain law of change (page 108). Does such a law hold good for hot and cold? By experiments with the spot apparatus mentioned above I was able to prove that it did ; the smallest noticeable change depends on the rate of change. But that complicated apparatus is not necessary to illustrate the law ; anybody can do it by means of a lamp and a spoon. Let some one else hold the spoon

An experiment.

by the extreme end ; you yourself put your finger about half way down the handle. The bowl of the spoon is now held over a lamp so that it will slowly become hot. If the lamp shines too hotly on your hand, you can put a screen in front. After a while the handle of the spoon under your finger begins to feel slightly warm. Lift the finger and immediately place the same finger of the other hand on the same place. The spoon will be found to be quite warm or even painfully hot. When the heat was *gradually* increased it was scarcely noticed, but when suddenly increased it was clear at once ; in short, the sensitiveness to heat depends on the *rate of change*.

At a very slow rate.

Although a frog jumps readily when put in warm water, yet a frog can be boiled without a movement, if the water is heated slowly enough. In one experiment (Fig. 70) the water was heated at the rate of 0.002°C ($\frac{36}{10000}$ of a degree Fahrenheit) per second ; the frog never moved and at the end of two and one half hours was found dead. He had evidently been boiled without noticing it.

It seems very strange that this law of mental life should

Importance of the law.

have remained unnoticed so long. In mechanics we study the velocity of a point; this would correspond to the rate of change in sensation. Physiologists have proven that in experimenting on nerves and muscles the effect depends on the rate of change. From psychological writers we have heard it repeated *ad nauseam* that there is no consciousness without change. What a little step it is to the statement that our appreciation of a change depends on the *rate* of change!

Example.

After all, every one of my readers has discovered the law already. "Why, how tall you have grown since I last saw you!" exclaims the visitor who has not seen Robbie for three months. "Do you really think so?" asks the mother. "I had not noticed it." The visitor had kept in mind Robbie's picture as she last saw him, and the change to the real Robbie of the present was sudden. To the mother the change had been gradual.

Fig. 70. Boiling a Frog without His Knowing it. No Sensation with an Extremely Slow Rate of Change.

Temperature and pressure.

There is a curious connection between temperature and pressure. Cold or hot bodies feel heavier than bodies of equal weight at the temperature of the skin. For cold, take two silver dollars; keep one of them closed in the hand to give it the temperature of the skin, but cool the other. Apply them in succession to the palm of some one's hand. The cold one will seem much heavier, which suggests a pleasant means of illusion for the poor man. Heat does not make so much difference as cold. For a successful experiment take two wooden cylinders of equal weight and heat one very hot in an oven. Apply the cylinders on end to the back of the hand.

CHAPTER X.

SMELL AND TASTE.

IN SPITE of the antiquity of language we have no names for smells. When we notice an odor, we name it by the source from which it comes. We speak of the odor of violets, of new-mown hay, of onions, and so on, but we have no name for the odor itself. **No names for smells.**

The structure of the smell organs in the nose has been studied most minutely and accurately ; their anatomy, as the science is called, is well known. The chemist can tell us accurately concerning most of the bodies from which we obtain smells. Strange as it may seem, the facts that interest us most of all, the smells themselves, have been neglected by science.

The lack of names for odors is very curious, especially because such a lack is not present in sight, hearing, or even taste. We might say that certain things taste like sugar, certain others like quinine, and so on ; but that would be only a roundabout way of saying they were "sweet" or "sour." Instead of classifying the colors, as grass-color, dandelion-color, coal-color, etc., we say green, yellow, black, etc. But in smell we can only speak of cabbage-odor, fishy-odor, violet-odor, and the like, for the language lacks names entirely.

Not only do we have no names for odors ; we do not know any reason why different things smell alike. Why should compounds of arsenic smell like garlic? If we **No reason for resemblances.**

mix sulphuric acid with water, we get an odor like musk. It is said that emeralds, rubies, and pearls, if ground together for a long time, give out an odor like that of violets. Again, ringworm of the scalp, the body of a patient sick with typhus, and a mouse have similar odors.

Groups of perfumes.

Perfumes can often be placed in similar groups. The rose type includes geranium, eglantine, and violet-ebony; the jasmine type, lily of the valley and ylang-ylang; the orange type, acacia, syringa, and orange-flower; the vanilla type, balsam of Peru, benzoin, storax, tonka bean (usually sold for vanilla extract), and heliotrope; the lavender type, thyme and marjoram; the mint type, peppermint, balsam, and sage; the musk type, musk and amber seed; the fruity type, pear, apple, pine-apple, and quince.

Threshold of smell.

What is the threshold of smell? There is a convenient but not highly accurate way of answering the question by means of what I shall call the "smeller" (olfactometer, or smell-measurer).

Olfactometer.

The smeller includes a glass tube (Fig. 71) fastened on a narrow board. Inside this tube is a narrow strip of blotting-paper moistened with the object to be smelled. A solution

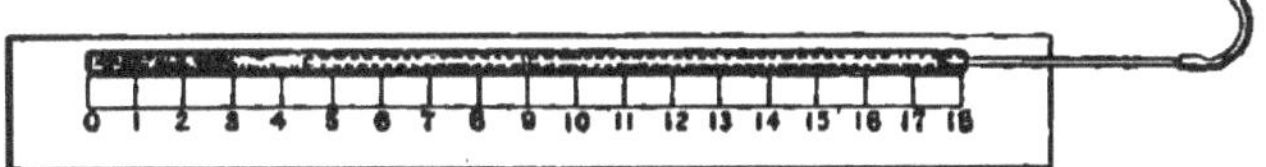

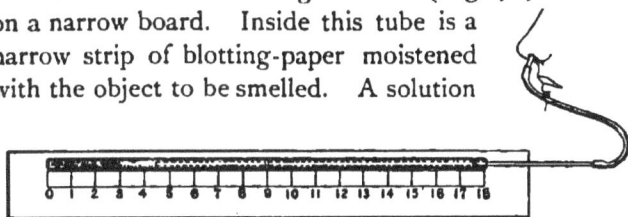

Fig. 71. Olfactometer, or Smell-measurer.

of camphor in alcohol is convenient; the solution dries, leaving the strip filled with small particles of camphor. Any other not too odorous liquid may take the place of the camphor solution. Inside this tube is a smaller one on the end of which is a piece of rubber tubing. A scale is marked on the board below the tubes.

Experiment.

The end of the smaller tube is pushed to the end of the larger one. The old air in it is blown out. The rubber tube is put to the nose. The smaller tube is now slowly drawn backward, while the person breathes in air through it. When he first perceives an odor, the distance through which the smaller tube has been drawn from the end of the larger one, is noted. Now, the further the tube is drawn back, the greater the distance over the blotter traveled by the air breathed; consequently there is more of the camphor odor in the air.

The number thus noted down gives an idea, though not a very accurate one, of the person's threshold of smell.

The threshold of smell will often be found to be different for the two nostrils.

Different thresholds of knowledge.

In the whole range of psychology there is nowhere to be found a more striking method of illustrating the difference between the different thresholds of knowledge. As the smelling-tube is pulled backward the observer at first notices no odor; the odor is said to be below the threshold. After a while he says, "I smell something, but I can't tell what it is"; a sensation is there, it is known as an odor; it has passed the threshold of sensation but has not reached the threshold of recognition (if I may use such an expression). The odor becomes stronger and stronger; finally the observer exclaims, "Now I know the odor; let me think a moment and I will tell you the name." Very frequently he recognizes the odor without being able to recollect the name. The difference between the threshold of sensation and the threshold of recognition is often considerable. If the odor is still further increased, the name, for usual substances, is readily recollected.

Fatigue.

Our sense of smell can be fatigued. Holding a piece of camphor for some minutes before the nose will raise the threshold for camphor. With an olfactometer charged with camphor the threshold as measured before fatiguing the sense of smell will be found to be much lower than the threshold afterwards. Sometimes the fatigue is so great that the smell of camphor is entirely lost. The cook soon ceases to notice the boiling cabbage, which appears so very odorous to a person just entering the house. The only way to live with people who eat onions or garlic is to eat them yourself; in a double sense, when in Rome, do as the Romans do.

Strangely enough the fatigue affects some odors and not others. If the sense be affected by camphor-fatigue, the smell of wax will be diminished or lost, but essence of cloves will appear undiminished in strength.

A whole laboratory can be found in the garden and in the pantry—a laboratory that has hitherto been put to little use for psychology. I suggest to my readers that they try the effect of fatigue of the odor of one flower, say tuberose, on the odors of other flowers.

Two senses of smell.

We have two senses of smell, the two halves of the nose. As it is difficult to attend to two things at the same time, it is but natural to expect that we cannot attend to both halves. Such is the case. When two different smells are received, one from each organ, we are driven to notice first one, then the other. When roses and water-lilies are both present we smell the combination of both ; but when a rose is placed in one paper tube and a water-lily in another and the tubes are so arranged that the odors get to separate nostrils without mixing, we do not smell a combination, but alternately either rose or water-lily. We can smell either one in preference to the other by simply thinking about it. It is a very curious fact that we are unable to think of the same odor steadily; our thoughts irresistibly turn from one to the other and thus the smells alternate.

Fig. 72. Alternation of Odors; or the Strife of the Two Nostrils.

Experiment.

Such experiments are possible to every one by use of paper cones. A sheet of paper is rolled into the shape of a candy-horn; the small end is trimmed off to fit the nostril. The flower is placed before the large end (Fig. 72). Odorous substances (perfumes) placed in bottles under the large ends can be compared in a similar manner.

Dullness and acuteness of smell.

The greater attention paid to sight and hearing has apparently caused a neglect of smell and a consequent deterioration. The acuteness of smell among animals is well known. Among certain persons this sense also attains great development. I have a case—reported by a perfectly competent witness who lived for years with the person mentioned—of a woman in charge of a boarding school who always sorted the boys' linen, after the wash, by the odors alone.

Names for tastes.

For the tastes we are much better off than for the smells; we have names for them. We say that something is "sour," that it is "sweet," etc., and do not need to name the taste after the object.

Flavors due to smell.

The great diversity of flavors of objects is due mainly to smell. When a cold in the head injures the ability to smell, the flavors of the dinner-table lose their value.

Experiments on taste without smell can be made by filling the nose cavity with water while the head is in an inverted position; simply holding the nose without breathing is almost as good.

Loss of smell.

When the sense of smell is entirely lost the ordinary flavoring syrups, such as vanilla, currant, orange, strawberry, and raspberry, give merely a sweetish taste with no distinction among them. Lemon syrup tastes sweet and sour. Candies flavored in this way taste alike. Mustard and pepper produce sharp sensations on the tongue; there is no difference between them except that pepper is sharper; neither produces a real taste. Tea does not differ from water or coffee, Rhine wine from diluted vinegar. Ginger and cloves are alike. Powdered cinnamon, when placed on the tongue of a person whose eyes are closed and whose nose is held between the fingers, is considered to be like meal.

Flavor of wines.

Wines owe their bouquet entirely to smell. The most exquisite Schloss Johannisberger does not differ from diluted vinegar as far as taste goes. The wines of Eastern Prussia are reputed to be at present good for nothing but to make vinegar, whereas in olden times they were considered good. This has been used as an argument to prove that the climate has changed; a much simpler explanation is that the early Prussians, owing to defective development of the sense of smell, did not know the differences among good wine, poor wine, and vinegar.

Coffee likewise owes its flavor to smell. Boiled coffee

has lost its aroma and is merely a combination of sour and bitter. Through unpardonable stupidity pepper is always served ground and consequently odorless, the little German pepper-mills being unknown in America.

Aromas.

Classification of tastes.

When all smells and touch and temperature sensations are gotten rid of, the things we taste can be sorted into six different classes: sour, sweet, salt, bitter, metallic, alkaline, and their combinations. Characteristic examples of these are found in lemon juice, sugar, salt, quinine, zinc, and washing soda.

Combination of tastes.

This does not mean that we experience only those six tastes. The elementary tastes can be combined in countless ways. Thus, sweet and sour when combined produce a result that is neither sweet nor sour, but different from either while resembling both. Unfortunately psychologists have not attempted to unravel the compound tastes into their elements.

Threshold.

Probably no more convenient or striking illustration of the threshold can be presented than in experiments on taste.

Threshold of sweetness.

The threshold for sweetness can be found by using a solution of sugar of known strength. An ounce of sugar dissolved in twenty ounces of water makes a five per cent solution. For simple illustration it is sufficient to place a spoonful of sugar in a small wine-glass of water. Some pure drinking water and two medicine droppers are to be provided.

A small glass is used, preferably a graduated medicine glass, containing one ounce of pure water. With one of the droppers a quantity of the sugar solution is taken up; one drop is allowed to fall into the water. The water is then stirred with the other dropper; a small quantity is taken up in it and one drop of this homeopathic solution is allowed to fall on the tongue of the person tested. He will not taste anything, owing to the extreme dilution. The experiment is repeated, adding one drop each time, till a taste is noticed. The number of drops used will indicate the threshold of taste. If the five per cent solution and a graduated glass have been used, it is an easy matter to calculate just how strong this least noticeable taste is.

Other thresholds.

Similar experiments can be made with salt, quinine, acid, and alkali. The most convenient solutions to use are sugar, 5 per cent; quinine, 0.002 per cent; tartaric acid, 0.5 per cent; salt, 2 per cent; sodium carbonate, 0.1 per cent.

To avoid the effect of suggestion it is advisable to have more than one solution ready and not to let the person tested know which is being used.

How far the education of the sense of smell can be carried is shown by the tea-tasters, who can tell the locality from which each chest of tea comes.

Intensity dependent on quantity.

Our appreciation of a taste depends on its quantity. A single drop of sugar solution on the tongue does not seem so sweet as a mouthful.

Intensification of one taste by another.

It is a very curious fact that a weak sensation of taste of one kind can be made to strengthen a taste of another kind. If two glasses of water be equally sweetened, one of them can be made to appear sweeter by adding a minute quantity of quinine powder. This is a fact which the cook must not make use of. The only other similar case that I know of occurs in hearing. Some partially deaf persons can hear much better in the midst of a noise. This is usually explained on purely physiological grounds, but there is a possibility of an explanation on the ground that the mind would naturally lump in a very weak sensation of any kind with the stronger one.

Taste with touch.

Some of the peculiarities of flavors are due to feelings of touch on the tongue. Soda water and champagne stimulate the tongue by the fine bubbles that they give off. Pepper and mustard produce an agreeable irritation. Puckery substances, such as raw quinces, act as the name implies. All such touch sensations are not tastes, although they and the smells enter into the flavors of things.

Sourness and touch.

Sour tastes are accompanied by touch. This can be brought out clearly in a series of experiments. We begin with a very weak solution of the acid, so weak that it seems like water when tasted. As it is made a trifle stronger, first a slight puckery feeling appears, even before the person experimented upon notices any sourness. By a little increase in the strength the sour taste is made to appear also. When the sour taste becomes very strong, a burning sensation is felt at the same time.

Saltiness and touch.

When we begin with a weak solution of salt and make it successively stronger, the taste appears first. Later a weak, burning sensation is felt ; this steadily increases but never overpowers the taste as in the case of sour things.

Sweetness and touch.

With a solution of sugar made steadily stronger a feeling of softness appears before the taste. Then the taste is most prominent. With a very strong solution we get the feelings of slipperiness and stickiness, as in honey and syrup. With saccharine (an intensely sweet substance) the touch sensations are present but not so prominent.

Bitterness and touch.

With bitter solutions made successively stronger, a fatty, smooth sensation appears before the taste. Thereafter the bitterness is most prominent. With pure quinine the bitterness overpowers everything, no matter how strong the solution. With quinine sulphate or chloride the very strong solutions are more or less burning.

Taste and temperature.

It is a curious but uninvestigated fact that temperature likewise has an influence.

Let equal quantities of water be placed in two tin cups, and let one cup be heated. Then if the same quantity of lemon juice or any sour solution be dropped into each, the warmer solution will taste sourer than the cooler one.

If a sweet solution be tried in the same way, the cooler solution will be the sweeter.

Now we can understand why housewives do not sweeten the rhubarb sauce till it is cool. If they sweeten it to taste while cooking, the acid taste will be stronger and much sugar will be required; when served cool on the table it will be too sweet.

CHAPTER XI.

HEARING.

AMONG the many sounds that we hear we generally make a classification into tones and noises. Pleasant sounds, like those of a flute, we call tones; unpleasant ones, like those of escaping steam, rumbling wagons, or screeching parrots, we call noises. This is only a convenient way of sorting sounds. Very many—if not most—sounds are either tones or noises according to the point of view. A jumble of piano-tones is a noise. The scraping of a violin produces a noise in the hands of a beginner and passes gradually from noise to tone as skill is acquired. A block of hard wood when struck makes a noise; yet we call the same sound a tone when the block of wood is one of the notes of a xylophone.

Tones and noises.

In a simple tone three properties are to be noticed: (1) pitch, (2) intensity, (3) duration.

Properties of tone.

As the finger is slid up or down the violin string, we hear changes in the pitch of the tone. As the bow is drawn harder or softer against the string, we hear changes in the intensity, or loudness. As the tone is continued for a longer or a shorter time, we hear changes in duration.

Pitch, intensity, duration.

We are so accustomed to saying that tones are "high" or "low," that there seems to be really something high or low about them. We might, however, just as well call the bass tones high. This naming of

"High" and "low" pitch.

the tones according to our notions of space is derived from the Middle Ages. The old Sanskrit terms meant "loud" and "soft"; the Hebrew was "audible" and "deep"; the Greek was "low" and "high" in exactly the opposite meaning to ours. The Latin was simply a translation of the Greek words for "acute" and "grave"; and the modern Romance languages, like the French, retain the Latin terms. In the Middle Ages it was customary to speak of ascending and descending; it is from this that German and English probably derive the highness and lowness of tones.

Extending the range of pitch.

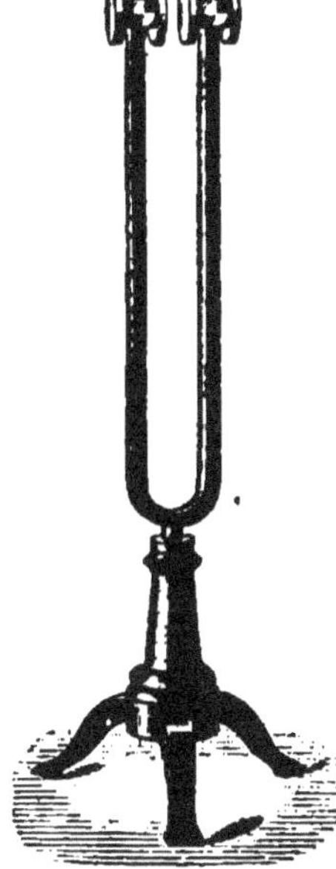

Fig. 73. Giant Fork for Finding the Lowest Audible Tone.

Starting from the middle of the piano, run the scale down toward the left. The lowest tone is very deep and shaky. Starting again, run the scale up to the right. The high tones sound shrill and tinkling. What would happen if the piano received lower and lower tones, or higher and higher tones, going on as long as we pleased?

Lowest tone.

To produce tones lower than the tones of the musical instruments gigantic tuning-forks over a yard long have been made. The way tuning-forks vibrate has been explained on page 104. The prongs are furnished with weights. As the weights are moved toward the ends the tone sinks lower and lower. In a short time weak puffs are heard in addition to the tone, each puff corresponding to a single movement of the prongs. Finally the tone disappears entirely, leaving nothing but puffs. The point at which the tone disappears is called the lower limit of pitch, or the threshold of pitch.

This lower limit is different for different persons. It is generally at about twelve complete vibrations per second. Some persons, however, have been found who cannot hear even the lower tones of the piano. Even the lowest tone of a large organ at thirty-two vibrations per second seems to some persons to be wavy.

Test for the highest tone.

Going upward in the scale we can proceed far beyond the piano. The test can be made with a set of small tuning-forks or small steel bars. It is most conveniently done with the Galton whistle (Fig. 74). This whistle can be altered in length by a screw-cap. As it is made shorter the tone rises. By means of a scale marked on the barrel the pitch of the tone can be calculated.

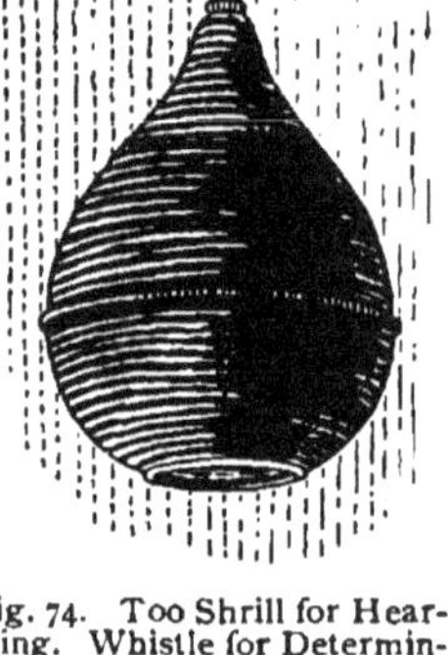

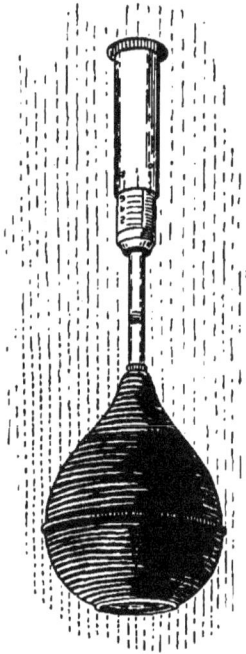

Fig. 74. Too Shrill for Hearing. Whistle for Determining the Highest Audible Tone.

The highest audible tone has been found to be very different for different persons. To some persons even the highest tones of the piano are silent. Others again can hear even up to 60,000 vibrations or more per second. The position of such a high tone would be musically indicated by the notation given in the margin.

Robert Franz, the composer of the music to Burns's "My Highland Lassie," in 1842 lost all the tones from E^3 upward in consequence of the whistle of a locomotive. In the following years he lost two half-tones more, so that in 1864 he heard nothing above D^3.*

The sound of a cricket is not heard by some persons. I cannot hear the squeak of a bat but believe, on authority, that it does make a sound. Many people cannot hear the shrill squeak of a mouse. When singing mice are exhibited, some people who go to hear them declare that they can hear nothing, others can hear barely something, and others again can hear much.

Effect of age.

It has also been noticed that as a person grows older he loses his power of hearing high tones. The persons themselves are quite unconscious of their deficiency so long as their ability to hear low tones remains unimpaired. It is an amusing experiment to test a party of persons of various ages, including some rather elderly and self-satisfied personages. They are indignant at being thought deficient in the power of hearing, yet the experiment quickly shows that they are absolutely deaf to shrill notes which the younger persons hear acutely, and they commonly betray much dislike to the discovery. Such persons should be comforted by the fact that every one has his limit. Sensitive flames have been found to be powerfully affected by vibrations that are too rapid for ordinary ears.

Limited range of high tones.

In some persons the upper limit of pitch is very low. It is related of Mr. Cowles, an American journalist,

---

*The reader is reminded that the successive octaves of the scale are indicated by small figures. Thus C^{-2}, C^{-1}, C^0, C^1, C^2, etc., indicate the successive C's of the scale; C^1 is middle C. The other notes are treated likewise.

that it was not until he was twenty-five years of age that he became perfectly cognizant of his defect. Up to this time he had treated all he read about the songs of birds as nothing more or less than poetical fiction. To him birds were perfectly mute; and he was perfectly deaf to the shrillest and highest notes of the piano, fife, or other musical instruments. At length, after considerable pains, he was convinced that he labored under some defect of hearing. When put to the test in a room where a large number of canary birds were singing very loudly, he declared he could not hear the slightest sound, even when placed close to their cages. Moreover, it was found that all the sibilant sounds of the human voice were equally inaudible. In all other respects his hearing was perfect.

Matters for speculation.

It is an interesting matter of speculation to consider all the tones we might still hear if our range of pitch extended higher. As a consolation we may remember what shrill sounds we now escape.

The question arises: When the whistle is too high for some persons to hear but not too high for others, does it produce tones? This I will leave as a nut to crack; "much can be said on both sides."

Highest tone for animals.

Galton, the inventor of the whistle, relates that he has gone through the whole of the Zoölogical Gardens, using a cane with a whistle at one end and a bulb at the other. He held it near the ears of the animals and when they were quite accustomed to the cane he would blow the whistle. Then if they pricked their ears it showed that they heard the whistle; if they did not it was probably inaudible to them. Of all creatures he found none superior to cats in hearing shrill sounds; cats, of course, have to deal with mice and find them out by their squealing. A cat that is at a very considerable distance can be made to turn its ear around by sounding a note that is inaudible to almost any human ear. Small dogs also hear very shrill notes, but large ones do not. At Bern, where there appear to be more large dogs lying idly about the streets than in any other large town in Europe, Galton tried his cane-whistle on them for hours together but could not find one that heard it. Nearly all the little dogs he met would turn around.

Influence of intensity.

Curiously enough the height to which we can hear depends on the strength of the sound. The results of specially made experiments are shown in Fig. 75. The figures at the bottom indicate the relative intensities of the blast of the whistle; thus the strongest tone used, 250, was five times as strong as 50, the weakest one. The figures at the left indicate the pitch of the highest audible tone for six different persons. At 50 for the person (F) the tone was lost at 10,000 vibrations, all above that being unheard. At 100 he heard to about 20,000; at 150 to 27,000, etc.

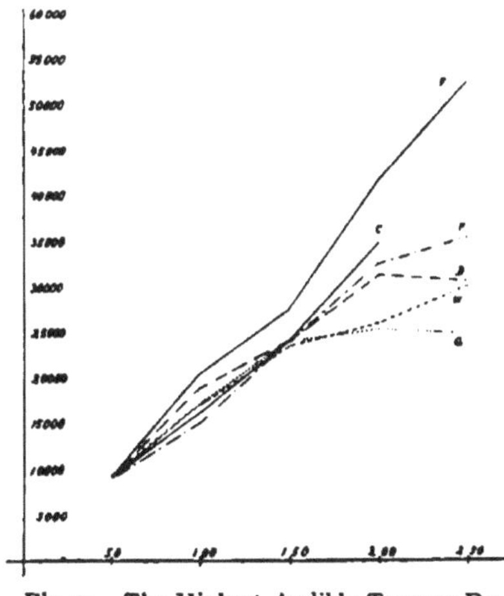

Fig. 75. The Highest Audible Tone as Dependent on Intensity.

Pitch is continuous.

Between the upper and lower limits of pitch the tones do not advance by steps as in the piano but continuously as in tuning a violin string. In other words, there is an unbroken range of tone, except in a few defective ears where portions of this range are lacking.

Least noticeable difference in pitch, or threshold of difference.

What is the least difference in pitch that can be noticed? Suppose that a violin is being tuned to another one or to a pitch-pipe, how nearly can we get it to an exact match? The fact that some persons cannot match tones as well as others is made plain by a few trials.

Apparatus therefor.

We wish, however, to get a measurement of the exactness to which we can judge tones, or, in other words, the accuracy with which differences between tones can be detected. This can be done by comparing a tuning-fork carrying an adjustable weight with one that remains always the same. As the weight is moved toward the ends of the prongs, the tone is lowered; as it is moved toward the stem, it is raised. Such a pair of forks is shown in Fig. 76.

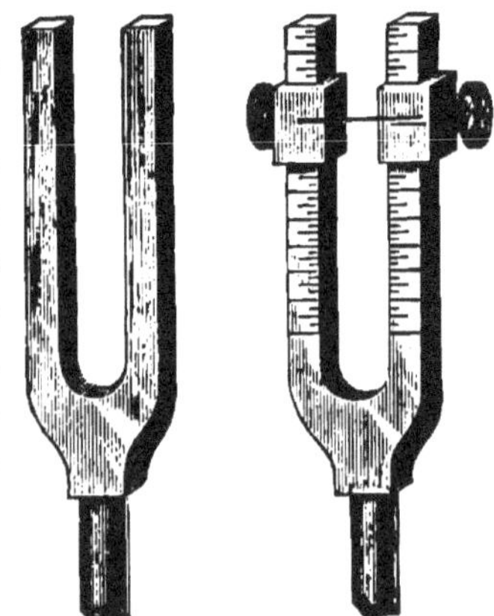

Fig. 76. Forks of Adjustable Pitch for Finding the Least Noticeable Difference.

Making the experiment.

The standard fork makes the same sound as the weighted fork when the weights are in the middle at 0. The standard fork is first sounded. Then after about three seconds the other is sounded. The person hearing them says at once whether he can detect a difference in pitch or not. If he says, No, the weights are moved a short distance toward the stem and the experiment is repeated. This is continued till he detects a difference, whereby the weighted fork is higher than the standard. This difference is called the least noticeable difference, or the threshold of difference.

Another way of experimenting.

Instead of a fork with adjustable weights a series of slightly differing forks can be used. To prepare such a series a dozen or more common tuning-forks all alike are obtained. The pitch of a fork can be raised by slightly filing the ends of the prongs ; it can be lowered by filing the prongs near the stem. Select one of the forks as the standard. Strike the standard and another fork at the same time, making them sound more loudly by resting them on the table or holding them opposite the two ears. If they are in the proper condition a single smooth tone will be heard. Now with a file slightly scrape the ends of the two prongs of the second fork, and sound them again.

Tuning the forks.

If the filing has been sufficient, the sound now heard will not be smooth and even, but will appear to wave between weak and loud ; often the forks will appear to say, "wow-u-wow-u-wow-u," etc. This peculiar effect is called a beat. It is known that the number of beats in one second is the same as the difference in the number of vibrations in one second. By counting the beats for four or five seconds the difference between the two forks can be readily determined. If the second fork is too high in pitch, it is filed more at the ends ; if it is too low, it is filed more at the stem. In this manner a whole set of forks can be obtained, differing by slight steps. For example, a convenient set is that of $A^1 = 435$ as a standard, with the other forks 436, 437, etc., as far as one has a mind to go. The preparation of such a series is somewhat laborious and, to fulfil all requirements, is somewhat expensive, owing to the large number of forks needed to provide for all ears from the finest to the coarsest. When the series is complete, the standard is compared with each in succession in the same way as with the adjustable fork until the just noticeably different fork is found.

Extension of the experiment.

Just as the threshold of difference is determined for a rise in pitch, so there is a threshold for a fall in pitch. The weights are started at the points where the two forks give the same tone. In successive experiments the weights are moved toward the prongs so that the tone of the weighted fork is repeatedly lowered. Finally the difference becomes noticeable. This is the point at the threshold of difference downward in pitch.

The tone-tester.

As there is some difficulty in finding out just what the pitch of the fork is for each position of the weights, and

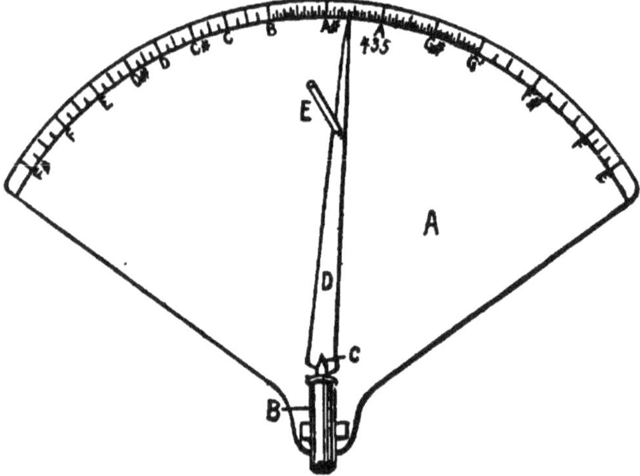

Fig. 77. The Tone-tester.

as the performance of these experiments takes a great deal of time, a more convenient instrument, called a tone-tester, has been devised. It consists of an adjustable pitch-pipe B fastened to a plate A. To the regulating rod C a long arm D is fastened, which is moved by the handle E. As C is moved inward, the tone of the pitch-pipe rises. As it is moved outward, the tone falls. Each movement makes a change in the position of the pointer. The tone-tester is compared beforehand with a carefully tuned piano to determine the position of the pointer when the pipe gives A of concert pitch. This position is marked at A 435 in the illustration. The figures mean that at this point the whistle makes a tone of 435 vibrations per second. In the same manner the succeeding notes are settled. The spaces are then subdivided by the eye into thirty-seconds of a tone.

Its use.

To make the experiment, the pointer is placed at A and the pipe is blown for an instant. The pointer is then moved upward one mark, and after about two seconds the pipe is again sounded. The person experimented upon tells if he hears a difference. The experiment is repeated, starting with A every time, till a difference is heard. In a similar manner the difference below A is found.

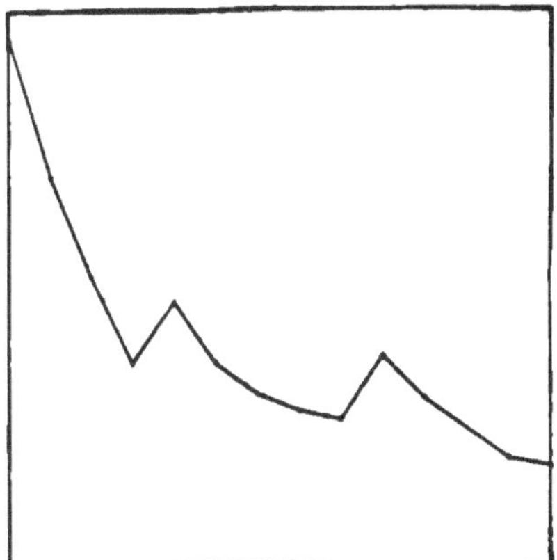

Fig. 78. Error in Hearing Decreases with Age from 6 years (at the left) to 19 years (at the right).

Experiments on school children.

In experiments made on a number of New Haven school children the accuracy in detecting differences was found to increase with age. The results are shown in Fig. 78. The distance along the bottom indicates the age, beginning at six and ending at nineteen. The distance upward indicates the number of thirty-seconds of a tone that could be detected. The smaller the number, the lower the irregular line and the more acute the child's ear.

Threshold of change.

There is another and perhaps more important threshold to be found than the threshold of difference, namely, the threshold of change. Almost all the experiments of psychologists have been confined to the threshold of difference ; I have lately called attention to this threshold of change and to the fact that it is an entirely different thing from the other.

The threshold of change can be illustrated by starting the tone at A and raising or lowering it continuously till a difference is noticed. I have succeeded in proving that the least perceptible change varies with the rate, as in temperature (p. 120), but have not been able to accurately determine the relation.

Can be determined for any tone.

These experiments give the thresholds only for A = 435. For the neighboring tones the pointer is started in the positions marked on the scale. For more distant tones other pitch-pipes would be needed.

Accuracy of tone-judgment.

There is another mental fact closely related to the tone-threshold but not quite identical with it, namely, the accuracy of tone-judgment. Suppose we have two forks almost but not quite alike in pitch. If we sound them in succession, we sometimes detect the difference, sometimes not. It is evident that for the same pair of forks the sharper ear will detect the difference more frequently than the duller ear.

Making the experiment.

The experiment is performed in the following way. Three forks are provided ; two of them are exactly alike, the third is slightly different. The person tested is seated with his back to the experimenter. The experimenter strikes two forks in succession; the person tested says at once whether they are the same or different. Suppose he says, Different; if they were really different the experimenter records one right answer. Suppose he says, Same; if they were really different the experimenter records one wrong answer. No record is made of the experiments with the two forks that are really the same, as they are introduced merely to avoid prejudice on the part of the person experimented upon. The experimenter finally counts up the total number of experiments with the two really different forks and the number of correct answers to these forks. For example, if there were twenty-five experiments in which the different forks were used and fifteen correct answers, the accuracy of judging this particular tone-difference can be stated for this particular person as $\frac{15}{25}$, or 60 per cent. With a greater difference between the two tones the percentage of correct answers will, of course, be greater. By using the same difference the relative accuracy for different persons can be ascertained.

Results.

The threshold differs greatly for different persons. Fine ears have been found that will detect a difference of less than half a vibration in tones between $B^0 = 120$ and $B^2 = 1,000$. Such observers can distinguish over 1,200 different tones within the octave B^1 to B^2.

Extreme cases.

On the other hand, it is not uncommon to meet persons who can hardly distinguish two neighboring tones. In fact, one case is reported of a well-educated man who had been unable to learn music in any way. It was found that he could not tell the difference between any two neighboring tones of the piano. Between the lowest tone and the highest he found a very great difference, but when the scale was run from one end to the other the change of tone appeared continuous and not by steps. In the middle regions of the scale he could not tell apart tones forming an interval less than a third; in the upper and lower regions the interval had to be a septime, an octave, or sometimes something still greater.

Finding the middle tone. If a low tone be sounded, then a mèdium one, and then a high one, we can tell whether the middle one is half way between the two extremes or not. Musical in-

Fig. 79. Apparatus for Finding the Middle Tone.

struments cannot well be used for this experiment as their tones are not simple but very complex; they introduce great errors into the result. By using tuning-forks perfectly pure tones are obtainable.

Apparatus. The arrangement for this experiment is shown in Fig. 79. Three tuning-forks, 1, 2, 3, are placed before adjustable boxes, or resonators, I., II., III. From each resonator a rubber tube leads to a general tube *s* which runs through double walls *z* to a distant room where the person experimented upon puts the end *o* to his ear.

In front of each box there is a movable cover which can be pulled aside by a string. Suppose the forks are sounding, the observer in the distant room hears nothing till one of these covers is pulled aside.

Fork 1 is selected as a low fork, fork 3 is selected as a higher one, and fork 2 is adjustable by weights. The forks are sounded in succession, 1, 2, 3 or 3, 2, 1. The observer tells whether fork 2 is properly adjusted to be in the middle or not.

Results.

The results indicate that our estimates do not follow the musical scale. For example, if the extreme tones be $C^1 = 256$ vibrations and $C^2 = 512$ vibrations, the middle chosen will on the average be $G^1 = 384$ vibrations. This is, counting by vibrations, just half way, but, according to our musical scale, it is nearer the upper tone. Likewise, if the extremes be $C^1 = 256$ and $C^3 = 1,024$, the middle will be about $C^2 = 840$ and not $C^2 = 512$.

Intensity of tones.

We have seen that everybody is deaf to very high tones and to very low tones. What about very weak tones?

Apparatus.

The first requirement is a tone whose intensity can be varied. This can be provided in many ways. The simplest plan is to use an electric tuning-fork in the manner shown in Fig. 80. A magnet between the prongs of the fork keeps it in motion electrically. The electric current is broken at every vibration of the fork. As it passes through the wire coil, it sets up electrical currents in the other wire coil near it. When a telephone is connected to this second coil, a tone can be heard by placing the telephone to the ear. This tone can be weakened by moving the second coil away from the first one.

Experiment.

The person to be tested puts the telephone to his ear. The second coil is placed far from the first; no sound is heard. It is gradually moved nearer till the tone is heard. Then it is placed close to the first coil, a loud tone being heard, and is gradually moved away till the tone is lost. The average of the two results gives a figure for the deafness of the person.

Crude experiments.

For rough tests a watch is often used. The watch is steadily brought nearer to one ear (the opposite one being closed) till the tone is heard. The distance of

Fig. 80. Everybody is Somewhat Deaf. Finding the Threshold of Intensity.

the watch from the ear indicates the threshold for sound, or the degree of deafness. This method is very unreliable, the chief difficulty being the disturbance by outside noises.

Illustration.

Probably no better illustration of this method of finding the threshold could be found than distant footsteps heard in a still night. All is silence. The assassin in his hiding-place feels secure from pursuit. Suddenly he notices a faint sound ; is it pursuit or imagination? It becomes louder and distinct enough to be clearly, though faintly heard; avenging justice is at hand. The intensity of the sound at the first hearing represents the threshold. The pursuers come nearer and nearer, but never think of searching the bushes by the wayside. Their footsteps die away in the distance; the last faint sound disappears at the threshold. Silence; escape at last.

The blessing of deafness.

We are, fortunately, all deaf. Every moving or vibrating object in this world would produce a sound to an ear sensitive enough to hear it. What should we do if our ears were so sensitive that the footsteps of every person between New York and California could be heard by a person in Chicago?

We are, unfortunately, not deaf enough to meet the demands of modern civilization. The incessant battery of noise and racket from rumbling wagons, factory whistles, car gongs, college clocks, clanging bells, house pianos, crowing roosters, whistling boys, and other diabolical inventions have been potent factors in producing what is known abroad as the American disease, neurasthenia, or nervous break-down. Until asphalt pavements, rubber tires, and laws against noise are introduced on this side of the Atlantic, there is no remedy but artificial deafness by stopping up the ears.

Notation for pitch and duration.

A special notation has been invented to indicate tones. The first complete notation for pitch is attributed to Guido Aretino in the eleventh century. Three centuries later the notation for duration was introduced by Jan de Meurs. Naturally the presence of exact means of expression for these two quantities afforded opportunity for progress in the artistic execution on the one hand and for scientific research on the other. The subject of pitch has reached a high degree of development. The duration of tones is also a matter of technique that has been carried to a great degree of precision in practice, although it has been scarcely investigated scientifically.

The staff notation.

We are all familiar with the staff notation for pitch and duration. For example,

indicates a certain tone of a definite character lasting through a definite time; it is the tone A with the length of one fourth of a whole note. By international agreement this tone has been fixed definitely so that it is the same in pitch throughout Europe and America; by a remark at the beginning of a piece of music the exact fraction of a minute occupied by a quarter-note is readily given.

Composer and performer.

Imagine the condition of music when the composer indicated the pitch but left the duration and time to the likings of the performer! Imagine the condition when the composer could indicate the pitch and the length of the tones but could not indicate their loudness or their form! I said "imagine" for the latter case; I meant to say "notice"—for that is the condition to-day.

Neglect of intensity.

The intensity of tones has been neglected; it must be remembered that we are not speaking of the semi-conscious use of the different degrees of intensity in the execution of a piece of music, but to a deliberate use of the shades of intensity. In music the consideration is confined to the five vague expressions, *ff*, *f*, *m*, *p*, *pp*. When a group of tones is to be made rather loud, put an *f* over it. How loud? just as the performer feels. All of the same loudness? just as the performer is in-

Helplessness of both composer and performer.

clined. Are all the tones without these letters to be of the same strength? just as the performer is disposed. These five vague grades cover only a few tones out of the thousands in a piece of music. The composer is powerless to give any indication of the wonderfully delicate shadings in the intensity of the different members of a group of tones; the performer is left without help. Two good performers on the organ will execute the same music with utterly different effects because they do as they please with the intensities of the tones. Which effect did the composer intend? Nobody knows.

Notation for intensity.

It is to overcome this difficulty that I propose a system of notes to include shades of intensity. Suppose, for the present, that we agree upon nine grades of in-

       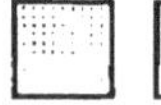 

Fig. 81. Method of Indicating Intensity in Notes; Loudest by Black, Weakest by White.

tensity between the weakest and the strongest the instrument is successfully capable of. Then we can introduce a system of shading to indicate grades of intensity just as the heraldist uses shading to indicate colors. Such a system is shown in Fig. 81.

Form of the tone.

This would cover the case in instruments like the piano, where there is no control over the form of the note. Most instruments, however, can produce tones of different forms. For example, suppose we are producing the tone

on the violin. We can make it steady in intensity from beginning to end; we can begin softly and go louder, or the reverse; or we could rise and sink in succession.

To indicate these differences we might use note-heads of the forms □ ◁ ▷ ⬭, where the first means a steady tone, the second means an increase from soft to loud, the third a decrease from loud to soft, and the fourth a rise and a fall in succession, or a crescendo.

The head of the note ought not to be used to indicate duration. In the present system duration is shown by the hooks on the stems of the notes, except in the case of the whole and half-notes, where a difference is made in the head of the note. This change in the head of the note is unnecessary for the indication of duration and can be employed to indicate intensity. A very slight change is thus necessary in the present notation ; we can retain the usual method of indicating pitch and the usual signs for duration with the exception of the two for the whole note and the half-note. These can be indicated by two lines across the stem of the ordinary quarter-note for the whole note and one for the half-note. Consequently the series of notes as regards duration will be that shown in Fig. 82, representing the whole, half, quarter, eighth, sixteenth, and thirty-second notes respectively.

Notation for duration.

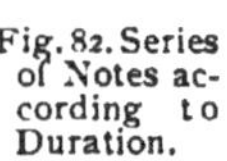

Fig. 82. Series of Notes according to Duration.

Whenever it is desired to write music without regard to intensity, it can be done in the same way as at present with the substitution of the two new signs for the whole and half-note, or it can be done as usual without any danger of there being a mistake in the playing of it. Moreover, the comprehension and the execution of pieces in the usual style will not be in the least interfered with.

No confusion with the old way.

Specimen notes indicating pitch, duration, intensity, and form.

Suppose we wish to indicate a half-note of medium intensity and even duration ; we have . Or an eighth-note of loud intensity and staccato form, . Or a whole note, weak but of crescendo form, .

Location of sounds.

Where are the tones we hear? With one ear closed the sounds we hear have no definite position. We know that a certain rattling must be down on the street because wagons cannot be up in the air ; the song of a bird cannot be under our feet. But a plain tone is nowhere, or rather, anywhere. Take a seat in this high-backed chair ; let some one hold your head firmly so that you cannot turn it. Put your finger tightly in one ear and close your eyes. Now I make clicks with a snapper sounder or I strike a glass with a spoon. Point to where the sound is. If I vary the intensity of the sound so that you cannot reason the matter out, your answers are generally wrong.

By turning the head you can get an idea of the place because you know that sounds straight out sidewise are stronger than in any other direction.

Judgment with both ears.

Open both ears but keep the eyes closed. Now you can tell me just where the sound is. You draw, unconsciously, an inference from the relative intensity of the two sounds from the two ears. But whenever I snap the sounder equally distant from the two ears, you are always wrong. Imagine a sheet of glass passed through the body dividing it into two halves symmetrically. For all sounds in this plane you are utterly at a loss. I snap my sounder under your chin ; you declare that it is behind your back. I snap it at your feet ; you say it is in front of your nose.

CHAPTER XII.

COLOR.

Color is a fact of mind.

"CONSIDER the lilies of the field ; Solomon in all his glory was not arrayed like one of these!" And yet flowers have no color, the rainbow has no color, all nature has no color, apart from the mind of the person seeing it. The flowers are beautifully colored to us because we see them. Those poor unfortunates who are totally "color-blind" see nothing but light and shade; those who are red-blind or green-blind see the world in mixtures of green and violet or red and violet; every one of us differs from every one else in his color-vision and sees the world in colors that differ for each person. The flowers have no colors; they send off physical vibrations, called vibrations of ether, but colors exist only where there are the eye and the mind to transform these vibrations.

Nothing to do with the physics of light.

With the physics of light we have nothing whatever to do except to provide apparatus for experiment; our problem is the study of color sensations. In the first place we shall treat the color sensations of the great majority of mankind. Those of my readers who are color-blind will, of course, soon find it impossible to understand what the rest of us are doing; they must wait for special attention.

The color-top.

A large amount of experimenting on the subject of color can be done by means of a properly selected package of colored papers (such as are used in the kindergarten) and a color-top or a color-wheel. The color-top was the invention of the great physicist Maxwell; it was used by Helmholtz for his investigations on color. There are numerous forms of the color-top; the miniature one shown in Fig. 83 has been prepared at my suggestion. The cost of these tops is so trifling that they can be given to school children by the thousand, like lead pencils or blank books, while at the same time the individual instruction thus obtained by every child makes the top more efficient than the color-wheel.

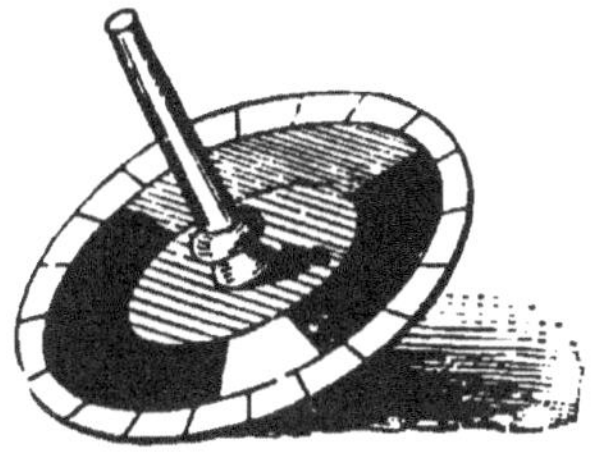

Fig. 83. The Color-top.

The color-wheel.

More convenient and accurate is the color-wheel, which has developed from the color-top. The best equipment of wheel and disks is that used by the physiologist Hering. Fig. 84 shows how the high speed of the disk is obtained by successive wheels.

Fig. 84. The Color-wheel.

The disks for top and wheel differ only in size. Each

disk has a hole exactly in the center to go on the axle; a slit runs straight from the edge to the hole.

Color-disks.

To put two disks of the same size together, they are slid over each other by means of the slits, as shown in Fig. 85. In Fig. 86 the two disks are shown ready to place on the axle. The proportions of the two colors can be changed at will by sliding one disk around on the other.

Method of use.

When the top is spun or the wheel is rotated, the different colors combine. If, for example, the red disk

Fig. 85. Putting Two Disks Together.

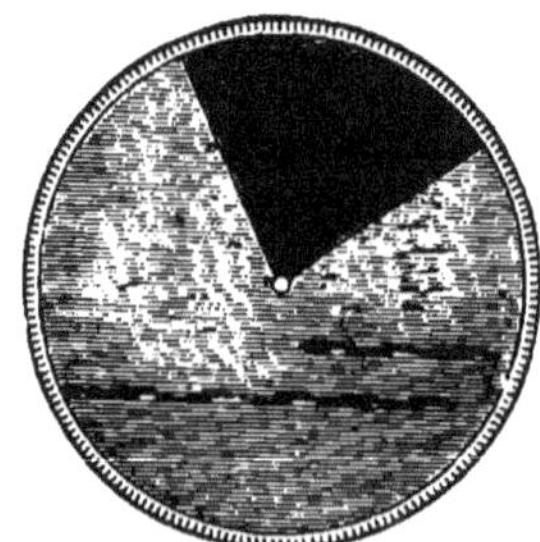

Fig. 86. Two Disks with Scale.

and the violet disk are placed together, the whole space will seem purple when the top is spun rapidly. If the disks, when still, show half red and half violet, the combined color will be a rich purple; if there is more red than violet, the purple will be reddish; and likewise the reverse.

Sorting colors.

The first experiment to be made begins with spreading the colored papers on the table. It is desirable to have a gray table-cloth. Choose any single color; place it on a clear space. Then place the paper next like it close beside it. Continue till all have been used.

Necessity of a system.

If the package of papers is complete, you will soon get into trouble unless you proceed methodically. Adopt as a principle that when two papers differ by being lighter or darker you will arrange them in a straight line with the darker end toward you.

Hue, shade, tint.

Suppose you have started with red ; then you will find, say, five reds alike except for their whiteness or darkness. Call them red, light red, very light red, dark red, very dark red. The bright red itself we will call a hue of red. The dark colors we can call shades of red, the lighter ones tints of red.

Succession of hues with the appropriate shades and tints.

Very similar to this red you will find a red that is slightly like orange, with all its shades and tints. This we will call orange red. Likewise you will find in succession reddish orange, orange, yellowish orange, orange yellow, yellow, greenish yellow, yellowish green, green, bluish green, greenish blue, blue, violet blue, bluish violet, violet, purplish violet, violet purple, purple, reddish purple, and purplish red. For each of these there are the appropriate shades and tints.

A color system.

Suppose you write the main colors in a horizontal line, as indicated in heavy type in Fig. 87. Now write all the tints above in smaller letters and all the shades below. Suppose that you find two tints passing off toward white ; for example, for red you have light red and very light red. You indicate them by LR and LLR. For the shades you have dark red and very dark red, DR and DDR. You will put OR beside R, LOR beside LR, DOR beside DR, etc. Continuing for all the colors, you get the complete plan in Fig. 87. For black we use D instead of B which would be confused with blue ; black is D-arkness.

But with purplish red you are only one step from red,

and the only way to bring it next to red is to cut the scheme out and bend it around into a cylinder.

Its final form, the color-cone.

But all the light colors, or tints, pass off toward white, and all the dark colors, or shades, converge

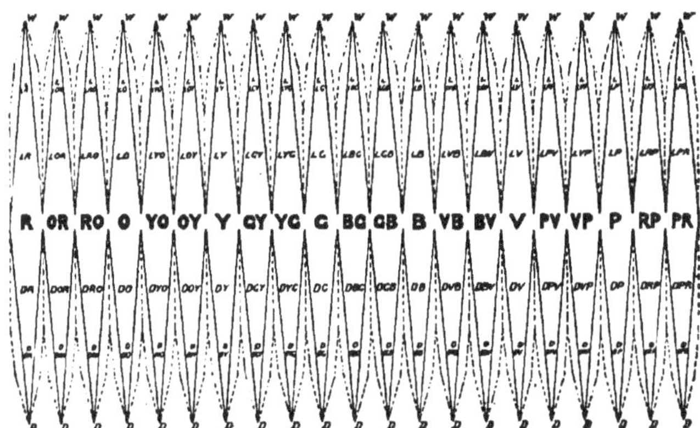

Fig. 87. Diagrammatic Arrangement of the Colors.

toward black ; they ought to be closer together than the other colors. By cutting out the figure along the zig-zag lines you can bring all the points together at the top and at the bottom. You will then have a double-pointed cone like that in Fig. 88.

The infinity of colors.

We are only finite, limited human beings and cannot even grasp the idea of the infinity, the unlimited number of full colors in the spectrum series. Red, for example, includes a large number of different reds passing gradually toward orange. We have divided them into red, orange red, reddish orange, and orange ; but we might just as well have made ten, twenty, or any number of subdivisions.

Tints are whitish colors.

The fact that tints are whitish colors is known to those who use paints. It can be proven by use of the color-top. Place together the red disk and the white one with almost no white showing; the resulting color is red. Add more white; the resulting color is a red tint. Add more and more white; the red passes through successively lighter tints till pure white is reached. The same is true of the other hues. Tints are mixtures of colors with white.

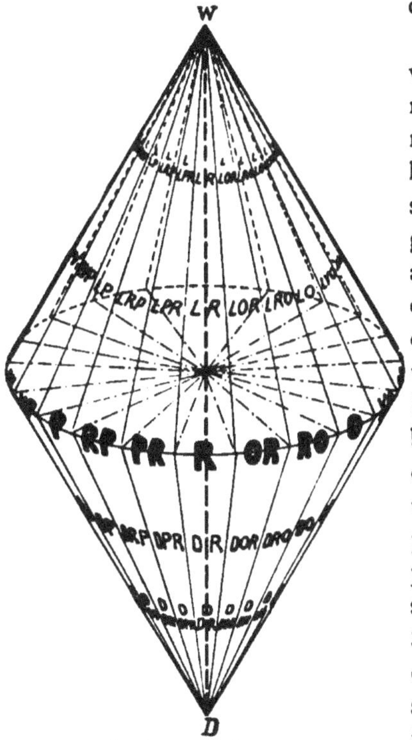

Shades are weaker colors.

The shades are weaker colors. Hold red and a shade of red squarely to the light. Keeping the shade in full light, gradually turn the red away so that it grows darker. At a certain degree of darkness it will match the shade. Place a red and a black disk on the color-top and gradually change their proportions. The red passes through all shades into black. Since black is absence of light, the red is simply decreased in intensity.

Fig. 88. The Color-cone.

With the top you can illustrate the fact that between a bright red and black, or between a bright red and white, there are countless intermediate shades and tints.

In our scheme of colors we have white and black, but no grays. If you take the color-cone in Fig. 88 and gradually cut off the light, the whole collection of colors becomes dimmer and dimmer till all, even the white, pass into black. In a dense night all colors are black.

Relation of colors to black.

Hold a piece of white paper squarely to the light; then gradually turn it away. It becomes darker and darker, grayer and grayer. Gray is, therefore, only darker white.

What is gray?

Take the same piece of paper into the sunlight. It is much whiter than before. What we thought was white was only a gray after all. Lay the paper on new-fallen snow. Alas! our whitest paper is a sorry gray when compared with God's white.

What is white?

The brightest and purest white is the light of the sun at noon on a clear day. All other whites are grays. We can therefore stretch a line from the whitest white to the blackest black and hang all the grays in between. Since the set is continuous and unbroken we will call it the set of grays, having white as the whitest gray and black as the blackest gray.

Absolute white.

White. Light Gray. Medium Gray. Dark Gray. Black.

FIG. 89. The Grays.

If we put a black and a white disk together on the color-top, we can imitate most of the grays by changing the proportions of the two. Of course, we cannot come anywhere near true white or true blackness.

What is known as the "absolute" white is the light of the sun at midday in a perfectly clear sky. The "standard" white for practical use is the color of magnesium oxide held in such sunlight; it is whiter than snow.

Standard white.

To produce the standard white take a piece of glass or mica and hold it over burning magnesium tape. Your friend the photographer is familiar with this process. His flash-light is almost as good as the magnesium tape. It is a good thing to keep such a standard of white; by comparison you will find that many white objects are tinged with red, yellow, blue, etc.

The line of grays passes through the middle of the color-cone from W to D.

Colored grays.

If we mix the colors with gray we get colored grays. This is readily done by using three disks on the color-top, a color, black, and white. Thus, red mixed with different proportions of black and white gives reddish grays, or, as our bric-a-brac friends would say, shades of terra-cotta. By using orange, we get the orange grays, or browns.

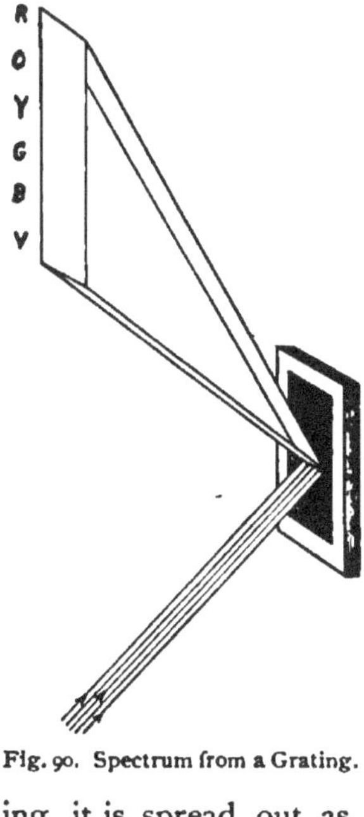

R
O
Y
G
B
V

Fig. 90. Spectrum from a Grating.

The brightest colors in nature.

The brightest hues to be found in nature are produced by allowing a ray of sunlight to fall on a spectrum-grating. This grating consists of a number of fine lines, 40,000 or more to the inch, carefully ruled on glass or metal. When a ray of sunlight falls on this grating, it is spread out as a band of color. By looking at the grating directly, the colors are seen without any contamination by reflection from objects. The band begins with violet and passes through all the colors of the rainbow to red. After the red comes the violet again, followed by the whole succession up to blue. After the orange comes purple, followed by blue, and so on. The band of color really consists of a series of rainbows ; the second and third overlap so that violet and red make purple ; the third and fourth overlap still more, and so on.

Standards of color.

The standards of color are found in the series from red to violet and in the purple. The standard colors from red to violet are also produced when a prism of glass is placed in a ray of sunlight. The colors from red to violet are called the spectrum colors. For the sake of convenience we sometimes add purple.

Nature's spectrum.

Nature uses the raindrops like prisms to form the rainbow. We may call the spectrum colors rainbow colors, but must remember that the great amount of white light in the sky mixes some white with them.

Artificial colors.

A moderately fair idea of some of the rainbow colors can be gotten from colored objects.

Red.

There is no pure red pigment in common use. The common idea of red is an orange red like vermilion. A very fair red may be obtained by mixing the pigments crimson-lake and vermilion. The deep ruby of the photographer's lantern is a very pure red. The red browns represent the shades of red. The common poppy is a beautiful red.

Orange.

Orange-peel is a very fair representative of orange. Red lead is orange with a slight mixture of red. Saffron is also a very fair orange. The shades of orange form the orange browns. The glow of a coal fire exhibits very fairly all the tints and shades of orange. The nasturtium is a characteristic orange.

Yellow.

An excellent example of yellow is found in pale chrome. Sulphur is a whitish yellow. Tan is a case of the yellow browns. The dandelion furnishes a good yellow; the buttercup is a whitish yellow.

Green.

Green is represented by the emerald green among paints. The greens of plants hardly approach the pure green. Apple leaves are nearly of the same hue, but they are much darker. Some of the greens in the plumage of tropical birds, especially of the parakeets, are a near approach to the standard green.

Blue.

Blue is represented by cyan-blue (Berlin blue) and ultramarine. A peacock's neck toward sunset is a very pure blue. As its flower we may choose the centaurea.

Violet.

One of the best examples of a pure violet is the color of the flower of some varieties of lobelia. The best time to see a pure violet is toward sunset. At this time the light from the sun is mostly violet, the red and other rays being weak.

In the spectrum thrown by the sunlight we find a great number of fine lines. The most prominent lines have received letter-names; thus, the two heavy lines close together in the yellow are called the D-lines. These lines are useful in defining the limits of groups of colors. Orange passes continuously through intermediate hues of orange and yellow into yellow; where shall we draw the line between orange and yellow? Helmholtz proposes the following system: red, all colors from the end to line C; orange, C to D; yellow, first quarter of the distance D to E; greenish yellow, from yellow to E; green, E to b; bluish green, b to F; blue, F to G; violet, G to end.

The flower spectrum.

Some of my fair readers may like to have a spectrum in the garden or on the flower-stand. For their special benefit I give a list of flowers for colors not too widely distant from the spectrum series.

THE FLOWER SPECTRUM.

Red.	*Orange.*	*Yellow.*	*Green.*
Poppy,	Nasturtium,	Dandelion,	Jack-in-the-Pulpit,
Cardinal Flower,	Chrysanthemum,	Lemon Lily,	Star of Bethlehem,
Tulip,	Tulip,	Tulip,	Cypripedium,
Celosia,	Azalea,	Primrose,	Cobia,
Geranium,	Marigold,	Marigold,	[Hydrangea],
Salvia.	Escholtzia.	Nasturtium.	[Mignonette].

Blue.	*Violet.*	*Purple.*
Larkspur,	Heliotrope,	Sweet Pea,
Cornflower,	Pansy,	Aster,
Forget-Me-Not,	Hyacinth,	Pansy,
Lobelia,	Crocus,	Phlox.
Flax,	Verbena,	
Centaurea.	Stocks.	

It is to be hoped that no interference with the rational system of naming colors will be allowed. In order to sell new wares the manufacturers are accustomed to invent new names for the colors, changing several of them every year. Some of the monstrosities thus perpetrated are "cadet blue," "crushed strawberry," "baby blue," "zulu," "ashes of roses," "elephant's breath," "calves' liver," "cerise," "gluten," "toreador," "eiffel," etc. I leave it to my readers to guess what the names mean. Monstrosities in color-names.

There is probably no more fascinating department of mental science than the study of the combination of colors. Newton was one of the first to show that the colors we see and their combinations have no counterpart in the physical world. As Maxwell states it, the science of color is a mental science. The little color-top puts into the hands of every one the power to make experiments on the most important laws of color combination. Combination of colors.

The first experiment is that of matching colors. Spread a piece of colored cambric on the table. Put a couple of the colored disks on your top and spin it on the cambric. Change the proportions till the top matches the cambric. Matching colors.

Now, note the number of hundredths of each color

A color equation.

shown on the top. Let x denote the color of the cambric and R, O, D, and W the colors of the disk you have used, D denoting black. Suppose you have 30 R, 45 O, 5 D, 20 W; then, since the cloth covers a whole circle, $100x = 30\text{ R} + 45\text{ O} + 5\text{ D} + 20\text{ W}$.

Practical applications.

How convenient for the forgetful business man! Madam wants a certain kind of brown trimming just like her piano-cover. She could pick out the right one by going to the store herself; she knows her husband will be sure to select wrongly, yet she cannot send a sample. So she spins the little top on the goods and adjusts the disks till she gets the proper brown. Now the man can put the top in his pocket and spin it on the store-counter till the salesgirl hands down the right color.

Or suppose a house-painter must order a new supply of color immediately. He matches it by his color-top and telegraphs the result. The dealer can at once adjust his own top and see the color wanted.

The necessity of such a method can be seen from the fact that the paints sold under the same name often differ widely. An English factory will produce a color that corresponds to 29 O + 71 Y while a German factory will make it 35 O + 45 Y + 20 D; and yet both colors will be called chrome yellow.

Fundamental color equations.

A color equation can be found for any color in terms of red, green, violet, white, and black. Pick up any piece of colored paper you find and cut a circle from it equal to the smaller disks of your top; cut out the center and slit it like the others. For the sake of brevity we will speak of this disk as x; all other small disks will be indicated by the small letters r, g, v, w, d, and the large ones by the capitals R, G, V, W, D. Put together the disk x and the black and white disks d and *w*. Likewise put together the large D, W, R, G, and V. Place the larger set on the top and then the smaller set over them. By repeatedly changing the proportions of the two sets of disks you can finally get them to match almost exactly. By help of the graduated circle on the top you can estimate the proportions of each color. For example, suppose a whole circle to be counted as 100 and the various colors to be in the proportion $17\ R + 45\ G + 10\ V + 28\ W = 54x + 46\ d$. Therefore, $54\ x = 17\ R + 45\ G + 10\ V + 28\ W - 46\ d$ or $x = \frac{17}{54}\ R + \frac{45}{54}\ G + \frac{10}{54}\ V + \frac{28}{54}\ W - \frac{46}{54}\ d$, which gives a definite color equation for the color of the paper.

Deductions from the experiments on combination.

The facts learned from the colored papers and the color-top will enable us to understand the laws of combination of colors. In considering this subject we have nothing to do with the physics of light or with the physiology of the eye, we must confine ourselves to mental facts just as we find them.

In the first place, between the two extremes, white and black, we have an unbroken line of neutral grays. Then we have a continuous line of colors according to hue, passing from, say, green through the blues, violets, purples, reds, oranges, yellows, back to green. Then by decreasing the intensity of these hues we can make each pass continuously into black, and by mixing each with any desired degree of gray or white we can make it pass continuously into gray or white. All of these can be produced by combinations of a few colors.

The infinity of colors can be produced by combinations of three fundamental colors.

This infinite number of colors of which we are capable can be produced from three fundamental colors, red, green, and violet. The fundamental red is nearly the same as a red that can be found in nature ; the fundamental green is a purer green than nature can exhibit; the fundamental violet is a natural color. If a beam of sunlight is reflected from a grating, it is broken up into a band of color like the rainbow. At one end is a pure red region, at the other is a pure violet region; these are two fundamental colors. The green of the spectrum is whitish. Since all nature receives its light from the sun, the colors of nature are limited by the character of sunlight. With the sun we have at present and are likely to have for the future, nature with all her colors does not give us all the greens we are capable of experiencing.

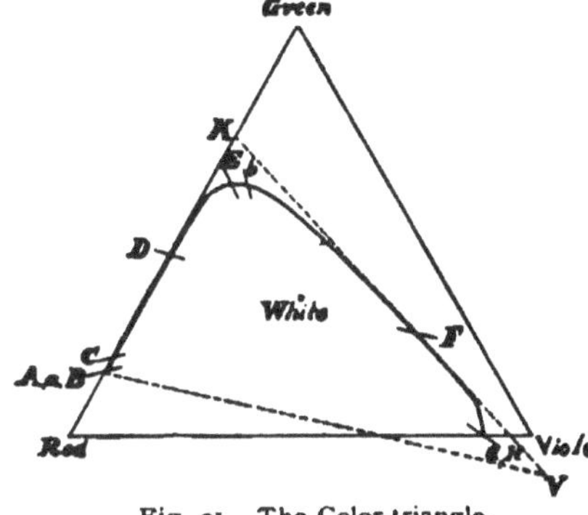

Fig. 91. The Color-triangle.

The color-triangle.

If we suppose our three fundamental colors placed at three corners of an equal-sided triangle with white in the center, the colors of nature would be enclosed within the curve drawn within it.

Properties of the spectral curve.

This curve has several remarkable properties. If we take any two colors, all the colors that can be produced by them will lie along the line connecting them. If we take spectral violet and spectral red, all the purples will lie along the line BG, the position being determined by the proportion of the two colors. If we take spectral yellow and spectral blue, the colors produced by mixing in various proportions will lie along the line DF, passing almost through white. If we wish to find what colors will produce white, we draw a straight line through white in every direction. Pairs of colors that produce white are called complementary colors. For three-color persons some of the simpler combinations are given in the table annexed. The table is taken from Helmholtz. It holds good only approximately, because the color-names are very indefinite ; thus, blue includes a group of blues which when combined with various members of the orange group give results varying more or less from white.

Complementary colors.

TABLE OF COLORS RESULTING FROM COMBINATIONS.

	Violet.	*Blue.*	*Blue Green.*
Red.	Purple.	Light pink.	White.
Orange.	Dark pink.	White.	Light yellow.
Yellow.	Light pink.	Light green.	Light green.
Yellow green.	White.	Light green.	Green.
Green.	Light blue.	Blue green.	
Blue green.	Water blue.		
Blue.	Indigo.		

	Green.	*Yellow Green.*	*Yellow.*
Red.	Light yellow.	Golden yellow.	Orange.
Orange.	Yellow.	Yellow.	
Yellow.	Yellow green.		
Yellow green.			
Green.			
Blue green.			
Blue.			

The combination of pigments, *e. g.*, paints, often gives a very different result from the combination of the colors directly. In fact, if the dyes with which two paper disks have been colored be mixed, a paper colored by the mixture will never be of the same color as the resultant from a direct mixture of the colors of the two disks by means of the color-top.

Combination of pigments.

This can be prettily illustrated by a disk prepared as in Fig. 92. The shaded portions are to be painted with blue, the light portions with yellow, and the central portion with a green formed by a mixture of half blue and half yellow. When the disk is rotated, the blue and yellow directly mixed never produce green but a grayish color with a blue or yellow cast.

When increasing quantities of yellow *paint* are mixed with blue paint, the color passes through various shades of bluish green, green, and yellowish green. When yellow and blue *colors* are mixed, the resulting color passes through grayish blue, gray, and grayish yellow. With some blues the gray has a very slight greenish tinge.

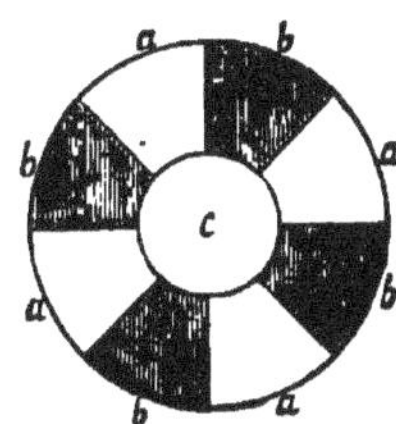

Fig. 92. Mixing Yellow and Blue.

Reason for the odd results in mixing pigments.

The reason why blue and yellow pigments give green can be illustrated by using blue and yellow glass. When two such pieces of glass are placed together, all light passing through both of them is green.

In colored glass.

Blue glass is blue because the glass absorbs the red, orange, and yellow light and allows the blue and violet light to pass. Yellow glass absorbs the blue and violet and allows the red, orange, and yellow to pass. Each of them allows a portion of the green to pass. When both of them are together, the blue keeps out the red, orange, and yellow, while the yellow keeps out the blue and violet. Consequently only the green gets through.

In paints.

Blue paints are blue because the minute particles of which they are composed send back to the eye mainly colors from the blue end of the rainbow series. Yellow paints send back mainly those from the red end. Both send back some green. When they are mixed, the blue paint absorbs all the red end and the yellow absorbs all the blue end, leaving only green to be sent back.

Similar results are obtained from the other paints; their mixtures are matters depending on their particular composition and not on their colors. Violet, for example, is one of the rainbow colors and cannot be produced by mixture of other colors. Yet red paint and blue paint can be made to produce a violet paint.

"Colors."

These accidents of the action of paints formerly led people to suppose that colors followed the same laws. Thus red, yellow, and blue were formerly called the fundamental colors. The artist often speaks of his paints as his "colors," and his laws of combination of the fundamental "colors" are quite correct, if by "colors" we understand paints. To avoid confusion with the other use of the word color, it is preferable not to use it to mean paint or pigment. Red, yellow, and blue are the fundamental pigments, and red, green, and violet are the fundamental colors.

CHAPTER XIII.

COLOR SENSITIVENESS.

Least noticeable difference in color.

WE ARE frequently called upon to distinguish small differences in color ; how accurately can we do it?

The color-top furnishes one method of answering the question. Suppose we take as a definite question : How accurately can we judge the mixture of small portions of blue with a large mass of red? The little red disk is placed in the center of the top ; it remains unchanged during the experiment. The large red and blue disks are placed together so that a minute portion of the blue appears. The top is spun ; no difference is detected. A little more blue is added and the top is again spun. This is repeated till the difference is noticed. The amount of blue can be measured by the graduated disk. Suppose it covers one half a space, that is, $\frac{10}{100}$ of the whole circle. The red must cover $\frac{90}{100}$, or nine times as much as the blue. Therefore we can add one part of blue to nine of red before the difference is detected.

The result depends upon the sensitiveness of the person. A dyer will detect minute differences that escape ordinary individuals; persons who have paid little attention to art are often incapable of detecting large differences.

According to age.

It has been proven that the sensitiveness to color differences increases with the age of school children. The results are given in Fig. 93. The figures at the bottom indicate the ages ; those at the side the relative amounts of difference that could be just detected. The greater the difference, the less the sensitiveness and the higher the curve. The steady descent of the curve shows the gain. In general, the girls were more sensitive than the boys.

Dependence on the intensity.

The sensitiveness depends upon the strength of the light. In very strong or very weak light it is much less than in moderate light.

The color of an object depends on the color of the

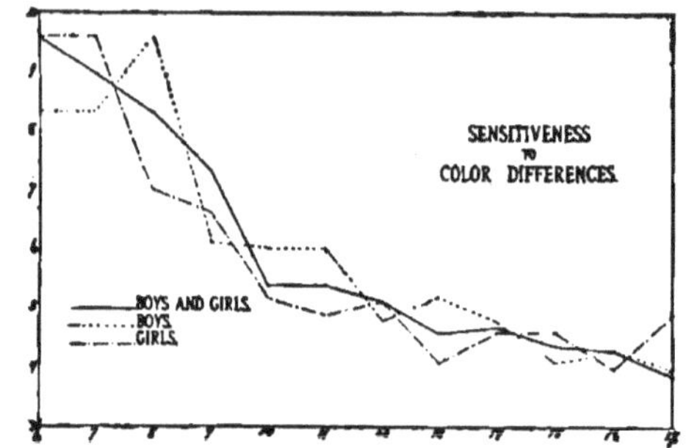

Fig. 93. Children have Finer Eyes for Color as they Grow Older.

The color of an object depends on its surroundings.

neighboring objects. If two designs are executed in the same gray, they will appear different if the grounds are of different colors. If the grounds are red and yellow respectively, one ornament will appear somewhat green and light, the other somewhat blue and dark. The effect is increased by placing tissue paper over them. Yet both grays are exactly alike. The color of the surrounding ground affects the gray.

Bits of gray paper laid on colored paper show the same result. If the colored paper be tipped so that the small piece slowly slides off, the colored tinge of the gray can be seen to slip off as the paper goes over the edge.

Contrast.

This influence of one color over another is called "contrast." The effect of the influence of a color is to spread the complementary color in its neighborhood. We have already defined as complementary those colors that give gray when mixed. White is complementary to black. Place a small piece of gray paper on a piece of colored paper. When a piece of tissue paper is laid over the whole, the gray paper appears distinctly colored. The tissue paper is used to whiten the colors, the contrast effect being greatest with whitish colors. Disks to illustrate contrast can be prepared as in Fig. 95, where the lined portion is to be colored. When such a disk is spun, the ring formed by the mixture of the black and white should be gray; it is, however, the color complementary to color on the other portion.

Effect of contrast.

Fig. 94. Disk to Illustrate the Effect of Contrast.

Plain red, violet, and blue woven cloths were on one occasion given to manufacturers for ornamentation with black patterns. When the goods were returned, the complaint was made that the patterns were not black; those on the red cloth were plainly greenish, those on the violet were dark greenish yellow, and those on the blue were copper-colored. By covering the cloth in such a way as to expose only the patterns without the colors, they were seen to be truly black. It was an effect of contrast.

Defective color-seeing.

When we started the subject of color, we agreed to talk for the majority and to attend to persons defective in color-seeing afterwards. Some of you may have already been puzzled by the apparent blunders the rest of us have been making. Let us see how we agree.

Detection of defective persons.

To measure the color-sense in different persons the two smaller disks, *w* and *d*, should be placed over the larger disks, R, G, and V. The white and black make a gray, and the larger disks should be adjusted to make a gray also. A finer adjustment is obtained by making both grays alike.

Fig. 95. Getting the Gray Equation.

The relative proportions of *w* and *d* may be disregarded and gray in general may be indicated by *m*. Suppose one person gets

$$m\ [= 60\ w + 40\ d] = 35\ R + 30\ G + 35\ V$$

and another

$$m = 5\ R + 45\ G + 50\ V.$$

It is evident that the second one is much less sensitive to red; in fact, such a person would be called red-blind.

Color-blindness.

Roughly speaking, humanity falls into four great classes: (1) the three-color; (2) the two-color red-blind; (3) the two-color green-blind; (4) the one-color persons.

Three-color persons.

The three-color persons form about ninety-five per cent of the males and almost all the females. They are so-called because the colors they see can be produced by combinations of three fundamental colors, red, green, and violet.

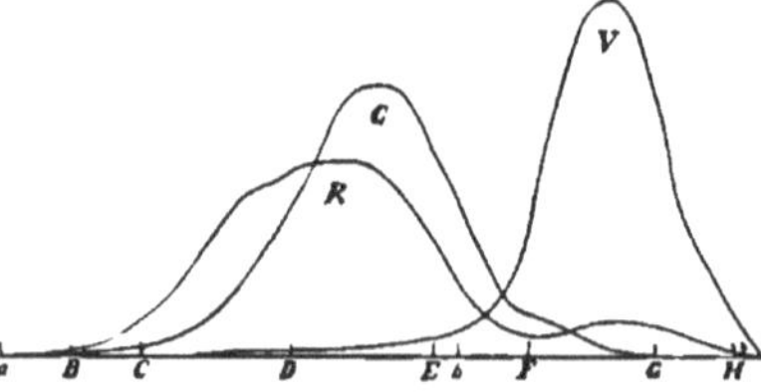

Fig. 96. Three-color Persons. Proportions of the Fundamental Colors in the Spectrum Colors.

Two-color persons.

The two-color persons form about five per cent of the males. The colors they see can be formed from two fundamental colors. If these two fundamental colors are green and violet, the person is said to be red-blind. If they are red and violet, he is called green-blind. The violet-blind persons are so rare as not to need notice.

Red-blind persons.

To the red-blind person red objects appear in general the same as dark green or greenish yellow; yellow and orange appear as dirty green; green is green, but is brighter than the false greens.

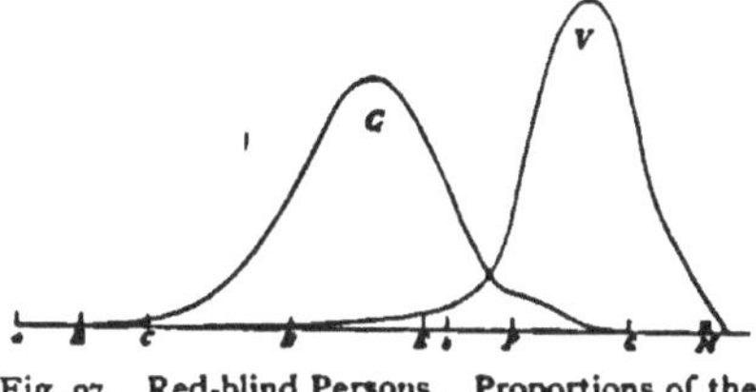

Fig. 97. Red-blind Persons. Proportions of the Fundamental Colors in the Spectrum Colors.

Green-blind persons.

The green-blind person calls red a dark yellow; yellow is called yellow but is lighter than the other; and green is called pale yellow.

Comparison with three-color persons.

The red-blind person is supposed to lack the fundamental red color. The colors he sees are all composed of green and violet; his color triangle (Fig. 91) shrinks up into a line GV, and his curve of spectral colors becomes a portion of this line. All colors of nature are to him mixtures of green and violet, gray (or white) being about the middle of the line. All the colors toward red are merely variations of green; all the purples are violets and blues. The green-blind person lacks the green; his range of colors is found along the line RV, with gray in the middle. The greens and yellows are reddish grays; the purples are also grays shading off into red or violet.

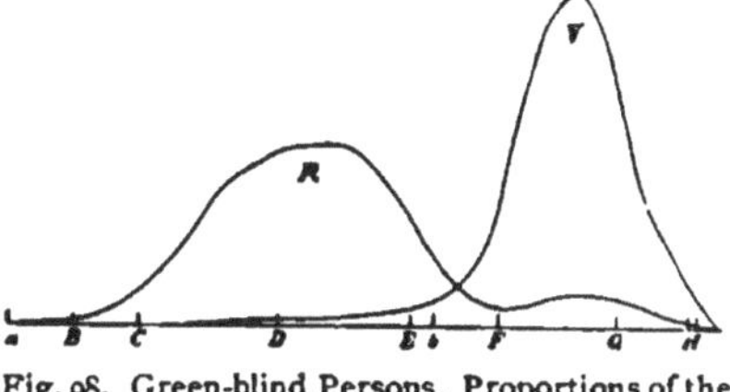

Fig. 98. Green-blind Persons. Proportions of the Fundamental Colors in the Spectrum Colors.

One-color persons.

The one-color persons see everything in light and shade, presumably gray. Their world is to the world of most people what a photograph or an engraving is to the radiance of nature. These persons are quite rare.

One case is related of an architect's assistant who did not understand in the least what was meant by color; he said that the colors appeared to him simply shades of white and black. He had to use colors in preparing the plans of buildings but was guided by the name on the paint. One of the clerks once purposely scraped off the names and he used the colors wrongly. A friend of his had a house with dark oaken timbers and light orange plaster. He asked, when looking at the house, why the plaster was so much darker than the wood. His friend told him that the plaster was very much lighter than the wood, but he refused to believe it. In a photograph which was afterwards taken the plaster came out much darker than the oaken timbers.

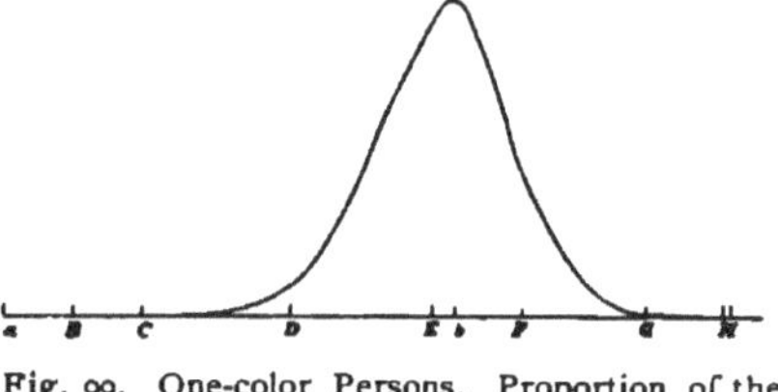

Fig. 99. One-color Persons. Proportion of the Fundamental Color in the Spectrum Colors.

The worsted test.

The phenomena of color-blindness are best studied with the color-top, but as the use of the top requires a great expense of time a number of quicker methods have been invented. One of the best is by use of the Holmgren wools. This consists of three skeins of worsted dyed with three standard test colors, namely, light green, pale purple, and bright red. Other skeins of reds, oranges, yellows, yellowish greens, pure greens, blue greens, violets, purples, pinks, browns, and grays are used as confusion colors.

Results.

The light green skein is laid before the person tested and he is told to pick out of the heap all colors that are like it. Nothing more is to be said; names of colors must not be used. If he picks out grays, brownish grays, yellows, orange, or faint pink, as the same, he is color-blind. Now the purple skein is laid before him. If he picks out blue or violet as the same he is red-blind; if he selects only gray or green he is green-blind. As a clincher, the red skein is used. A red-blind person will match this with dark greens or dark browns, while the green-blind person will choose light greens or light browns. Are we to suppose that the many Englishmen are color-blind who can see in the Irish flag only a symbol of anarchy?

Color-weakness.

Numerous modifications of this method of testing have been used. The method is not always successful when the person tested is not color-blind but color-weak. He may be able to pass the tests in a bright light, and yet he cannot distinguish red and green in a fog, or he may have perfect color-vision near by and be color-blind for objects at a distance.

Lantern test.

A lantern with colored glasses is sometimes employed. A color is shown to the person tested; he names it. Other colors and white are shown in succession. Then gray glasses to simulate fog are used over the colors, in order to detect the color-weak.

Red and green signals.

The matter of color-blindness has been brought into notice by the use of red and green lights as signals on railways and boats. Red means "danger"; green means "all right" on the railway. On the water red is the port side of the boat, green is starboard side; a pilot knows which way a vessel is sailing by seeing red or green. It is evident that any inability to distinguish them is a source of danger.

The steamship *Isaac Bell* collided with the tugboat *Lumberman* near Norfolk, Virginia; ten lives were lost. The pilot of the *Lumberman* was afterwards examined and found to be color-blind; there was a rumor that the other pilot was also color-blind.

Accidents due to color-blindness.

The pilot of the steamer *City of Austria*, which was lost in the harbor of Fernandina, Florida, was proved to be color-blind. He mistook the buoys, and his mistake cost the owners $200,000.

A case of color-blindness.

Captain Coburn reports: "The steamer *Neera* was on a voyage from Liverpool to Alexandria. One night shortly after passing Gibraltar, at about 10:30 p. m., I went on the bridge, which was then in charge of the third officer, and competent in every way. I walked up and down the bridge until about 11 p. m., when the third officer and I almost simultaneously saw a light about two points on the starboard bow. I at once saw it was a green light, and knew that no action was called for. To my surprise the third officer called out to the man at the wheel, 'Port,' which he was about to do, when I countermanded the order, and told him to steady his helm, which he did, and we passed the other steamer safely about half a mile apart. I at once asked the third officer why he had ported his helm to a green light on the starboard bow; but he insisted it was a red light which he had first seen. I tried him repeatedly after this, and although he sometimes gave a correct description of the color of the light, he was as often incorrect, and it was evidently all guesswork. On my return I applied to have him removed from the ship, as he was, in my opinion, quite unfit to have charge of the deck at night, and this application was granted. After this occurrence, I always, when taking a strange officer to sea, remained on the bridge with him at night until I had tested his ability to distinguish colors. I cannot imagine anything more dangerous or more likely to lead to fatal accidents than a color-blind man on a steamer's bridge."

Another case.

A similar account is given by Capt. Heasley, of Liverpool: "After passing through the Straits of Gibraltar, the second officer, who had charge of the deck, gave the order to port—much to my astonishment, for the lights to be seen about a point on the starboard bow were a masthead and green light; but he maintained that it was a masthead and red, and not until both ships were nearly abreast would he acknowledge his mistake. I may add that during the rest of the voyage I never saw him making the same mistake. As a practical seaman I consider that a great many accidents at sea arise from color-blindness."

Insufficiency of the test by wools.

The following is an extract from a letter by a "thirty years' railway man." "I have been on the railway for thirty years and I can tell you the card tests and wool tests are not a bit of good. Why, sir, I had a mate that passed them all, but we had to pitch into another train over it. He couldn't tell a red from a green light at night in a bit of a fog."

Color-blindness among the Quakers.

Color-blindness is hereditary. Among the Quakers, for example, the proportion of color-blind persons is about one half greater than among other people. Nearly every Quaker is descended on both sides solely from a group of men and women who separated themselves from the rest of the world five or six generations ago. One of their strongest opinions is that the fine arts are worldly snares; their most conspicuous practice is to dress in drabs. A born artist would never have consented to separate himself from the soul-stirring artistic productions of his fellow-men ; he would have felt that such an action would be treason to the instincts that God planted in him. It is quite probable that Quakerism would be very likely to attract to itself not only those who were lacking in instinct for the beautiful, but also those actually color-blind. The productions of many of our artists must appear actually hideous to color-blind persons who cannot tell the difference in color between a strawberry and its leaves. Again, the desertions from Quakerism would naturally be of persons in whom these instincts and abilities were stronger. Dalton, the discoverer of color-blindness, was a Quaker. It is related of a prominent Quaker that he returned from town one day with a bright red tie, a perfect abomination to his family. In spite of the trouble aroused, it was not a case of heresy but merely of color-blindness.

Explanation of Quaker customs.

Not heresy, but color-blindness.

CHAPTER XIV.

SEEING WITH ONE EYE.

The world seen with one eye.

LET us look at the world with only one eye. What we see consists of patches of color arranged in wonderfully complicated forms. It is our duty to determine some of the laws of this arrangement in space.

Point of regard.

The first fact that strikes us is that we are looking at some particular point. This is the "point of regard." In looking at this dot ● your point of regard is the dot. As you read onward, your point of regard changes from one letter to another. If you look at a person on the street, the point of regard is that person.

Distinct and indistinct vision.

Keeping the eye steadily looking at the dot, notice that you can read the words close around it although they are somewhat blurred, and that, although you can see over a whole region, including the page and part of the room, all this region is quite indistinct. The fairly clear part around the point of regard is the region of distinct vision; the blurred part is the region of indistinct vision. The whole region seen is called the field of vision.

Boundaries of the field of vision.

The boundaries of the field of vision are determined by moving objects from outside the field toward it until they are seen, and by moving them from the center outward till they disappear. The subject of experiment is seated in a chair; one eye is closed, the other looks, without moving, straight ahead at a spot. The experimenter places a small piece of white paper on the end of a knitting-needle or a stick and, starting behind the subject, slowly pushes it forward at about one foot from his head till he catches sight of it. The paper is then started where it is seen and is drawn back till it disappears. This marks the limit of vision in that direction.

Perimeter.

The limits of the field of vision are determined and recorded rapidly by means of perimeters. One form is that shown in Fig. 100. The small piece of paper is moved out along the curved arm in one direction till the limit is found. The arm is placed in various positions and the experiment is repeated. The number of degrees is read off each time and is marked on a chart. A line drawn through these points indicates the boundary of the field of vision. An average eye will have a field extending outward (*i. e.*, away from the nose) about 85°, inward 75°, upward 73°, downward 78°.

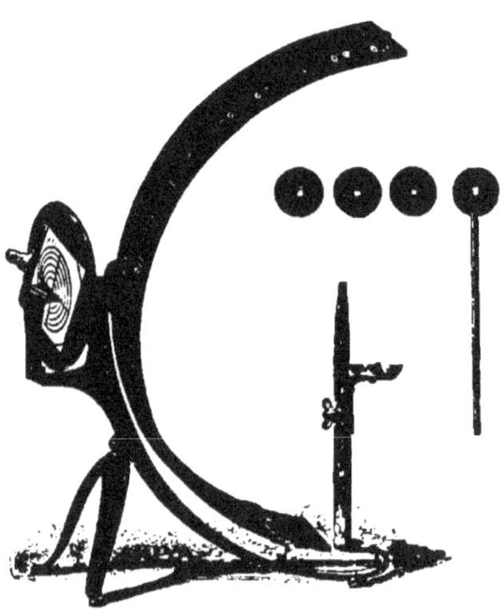

Fig. 100. Perimeter, for Measuring the Field of Vision.

Color-limits.

If the experiments on perimetry are made with colored objects, it will be found that in a narrow region along the edge of the field of vision the person can see the object without seeing its color. In fact, in this region we are all totally color-blind ; we see everything in an indefinite gray color.

It was formerly supposed that just inside the one-colored region there was a red-blind region and the rest of the field was evenly three-colored. Recent experiments indicate that the case is not so simple.

Inside this one-colored border the object takes on a color, but the color is seldom the same as that which it has when seen directly. The limits at which objects of

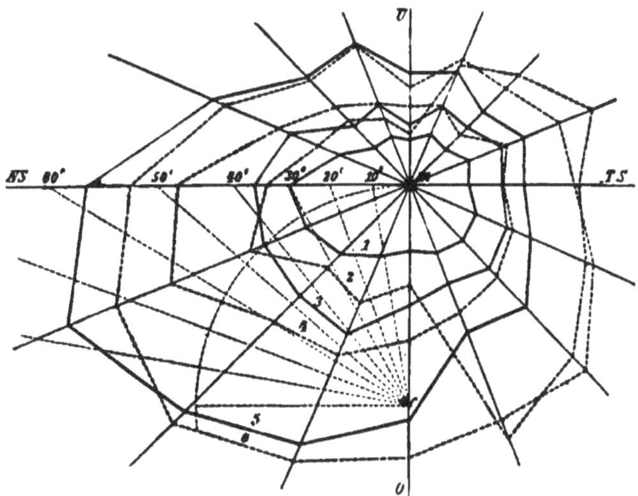

Fig. 101. Perimeter Chart. Limits beyond which the colors disappear: 1, Violet; 2, Yellow; 3, Green; 4, Red; 5, Orange; 6, Blue.

various colors lose their "true" colors, *i. e.*, the colors when directly seen, are indicated for a specimen person in Fig. 101.

Field of vision in children.

Curiously enough, the field of vision with children is apparently not so great as with adults. They cannot see over so much for any position of the eye. The reason probably is that they are incapable of attending to the outer regions ; they confine themselves to the region near the point of regard.

Blind-spot.

In the field of vision there is one place at which nothing is seen ; this is called the blind-spot.

How to find it.

With the left eye shut, hold the book at arm's length and look with the right eye at the cross in Fig. 102. The letters are also seen indirectly. Bring the book slowly toward you, keeping the eye fixed on the cross. Suddenly the B will disappear entirely. If the book is brought still closer the B will reappear, but the O will disappear, leaving a blank space between A and B. There is one portion of the field of vision on which you are absolutely blind.

+ A O B

Fig. 102.

To try the left eye, hold the book upside down.

Its discovery.

Although man and his animal ancestors have always had blind-spots as long as they have had eyes, these spots were not discovered till about two hundred years ago, when Mariotte caused a great sensation by showing people at the English court how to make royalty entirely disappear.

How to draw it.

The blind-spot can be drawn directly on paper by keeping the eye fixed on the cross while a pencil is moved from the circle outward till its point is just seen.

In this way a dotted boundary line for the spot is obtained.

Its size.

The blind-spot ordinarily covers a region equal to the face of a man seven feet distant, or eleven times the size of the full moon.

What is seen at the blind-spot.

What do you see at the blind-spot? Everything disappears that is put in the region covered by it. Yet there must be something there; for, if the O in Fig. 102 be made to disappear, the letters are no nearer together than when the circle is seen.

The blind-spot must be seen as white, for the whole region appears unbroken. Yet if this experiment is made on colored paper the whole region is of the same color. Papers or cards of various colors can be readily prepared to illustrate this. We are thus forced to the conclusion that although we are blind over this region, we fill out the lacking space by an unconscious act of imagination and that it is filled out in accordance with the surrounding region.

Puzzling the blind-spot.

Let us, however, try to puzzle the blind-spot. A card is prepared with colors as shown in Fig. 103. Let

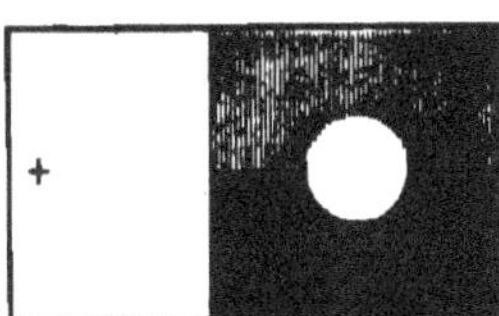

Fig. 103. Putting a White Circle on the Blind-Spot.

Fig. 104. The Circle is Replaced by the Colors.

the white circle fall on the blind-spot. The card will appear as in Fig. 104.

Try a card colored as in Fig. 105. If the circle falls on the blind-spot it will be filled out as in Fig. 106.

Now try a card like Fig. 107, with the circle brought into the blind region. At last the spot is puzzled. One moment the blue band will run across the red one; at another the red will run across the blue. Sometimes after many trials the spot seems to despair and the person owning it declares that he really sees nothing there.

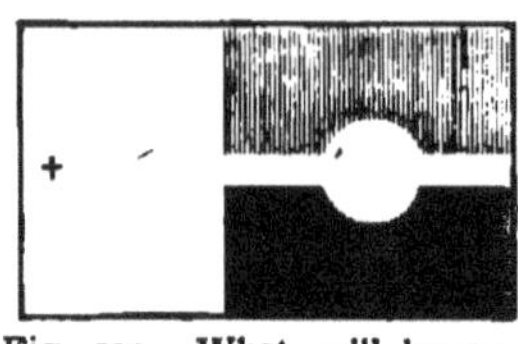

Fig. 105. What will happen now?

Deception by the blind-spot.

In looking at a printed page the portion that falls on the blind-spot appears to be printed with indistinct letters, as though it were pretending to read.

Fig. 106. The Result.

No influence on space.

It is noteworthy that the space around the blind-spot is not contracted. If the circle in Fig. 102 falls on the blind-spot, the letters A and B are no nearer together, although quite a space has apparently been removed.

Fig. 107. A Puzzler for the Blind-Spot.

Movement of the point of regard.

Up to this point nothing has been said of motion in connection with vision. We can move our point of regard at will. Indeed, the point of regard cannot be kept steadily on any object. Try to look steadily at the white dot in Fig. 109. You will soon see the edges of the white circle blurred over by the black edges. The point of regard trembles and sways like the pointer described on page 74. Some persons of nervous temperament cannot approach even a moderate degree of steadiness.

We have already noticed that the point of regard can be moved around in any direction. It is mainly by our knowledge of such movements that we judge the size of objects.

Difference in difficulty for different directions.

It is readily noticed that when the head is held upright and the point of regard is taken directly in

Although the blind-spot pretends that it can read and will attempt to deceive by making this space appear covered with letters, yet, if you look steadily at the cross (with the left eye closed) and place the book at such a distance that the black dot disappears, *i. e.*, falls on the blind-spot, you will find that the letters imagined by the blind-spot are only indefinite marks.

Fig. 108. The Blind-Spot Pretends to Read.

front, the upward movement is more difficult than the side movements or the downward movements. Let us measure these movements on each other.

Errors in estimating space.

Put a blank sheet of paper on a board and place a dot in the middle. Holding it directly in front of the eye so that the dot is at the point of regard looking straight forward, draw four equal lines, as indicated in Fig. 110. On measuring these lines the vertical one above the dot will be found shorter than the vertical one below. Both will be shorter than the horizontal lines; the horizontal lines will generally be equal. We can thus conclude that space above the point of regard in the usual position is overestimated as compared with space below; that space in a vertical direction is overestimated as compared with horizontal space; and that horizontal space inward or outward is about the same.

Fig. 109. Test for Eye-Steadiness.

This explains why *c* and not *b* seems the continuation of *a* in Fig. 111.

Apparent and true square.

Placing a dot on the paper in the same way, draw a square around it. By turning the square sidewise you will see that you have really made it too short. Turn this book upside down. What do you notice in regard to the letter s and the figure 8? Why are they made so?

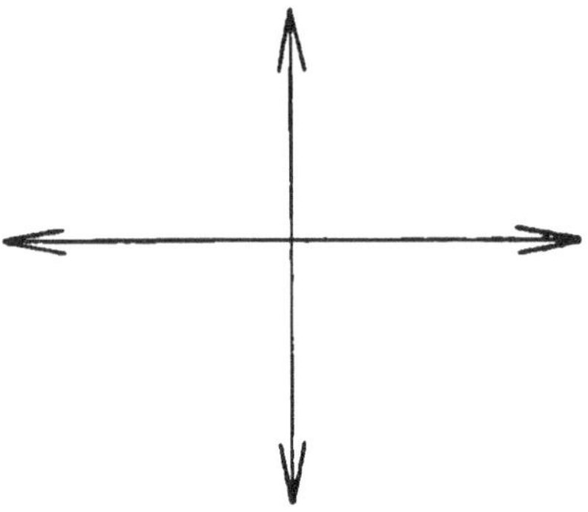

Fig. 110. What the Eye Considers to be Equal Distances.

Tipping of vertical lines.

When the point of regard moves upward it has a tendency to move outward; when it moves downward, it moves also inward.

Looking at the edge of the room, you will notice that as you look rapidly along it toward the ceiling the whole edge seems to tip inward. With the right eye it tips toward the left, with the left eye toward the right. As you look rapidly downward toward the floor, the edge appears to tip in the opposite direction.

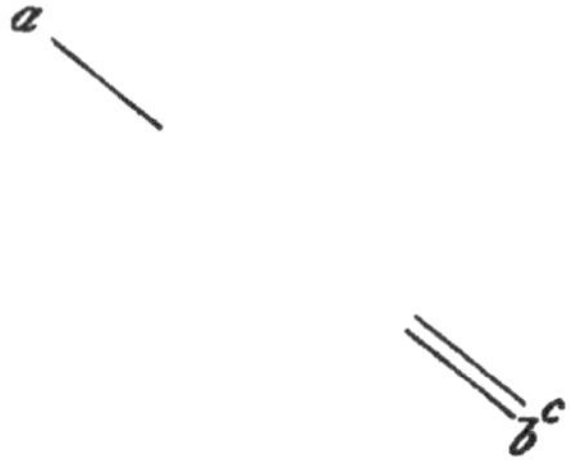

Fig. 111. Which is the continuation of *a*? Why?

This tipping is very disagreeable in the cities of tall buildings. If you happen to look at them from one side of the eye, they seem to be leaning dangerously over the street; if from the other, they seem to slant back as if disdaining the streets below them.

The amount of this tipping in the eye can be measured. Rule a horizontal line on a sheet of paper ; then lay the edge of the ruler across it at what you judge with one eye to be a right angle and draw the line. On another sheet of paper do the same for the other eye. Your two right angles will disagree to a small extent.

Illusions of distance.

Distances are judged by the difficulty in traversing

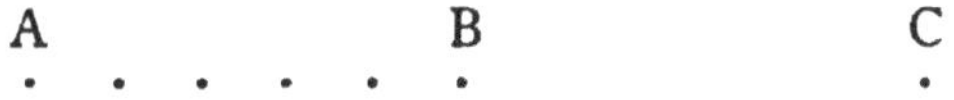

A B C

Fig. 112. Illusion of the Interrupted Distance.

them ; if the road is hard, or if you make many stops by the way, it is much longer than otherwise.

Interrupted distance.

The distance between the two dots A and B in Fig. 112 is apparently greater than that between B and C. The intervening dots are like tempting seats by the wayside.

Fig. 113. Illusion of Filled Space.

The journey is really made harder and apparently longer because your attention is caught at each one.

Constraint in movement.

The open distance in Fig. 113 is apparently less than the line-distance. It is harder to walk on a straight and narrow path than to go as you please; you may go perfectly straight anyway, but with no directing line you are free from constraint. This explains why the continuation of *a* in Figs. 114 and 115 appears to be at *c*.

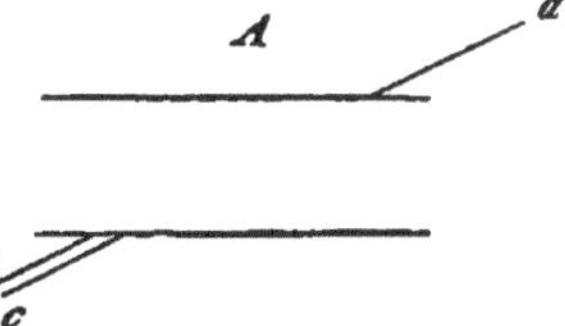

Fig. 114. Which is the continuation of *a*? Why?

We have learned to estimate distances by movements

of the point of regard, and the whole visual field is regulated accordingly. Even without movements we make the same errors of estimation.

Errors without movement.

The illusion in Fig. 116 is on the same principle as that in Fig. 113. The effect depends on the relation of the number of cross-lines to the distance; with too many or too few it is not so powerful.

Influence of cross-lines.

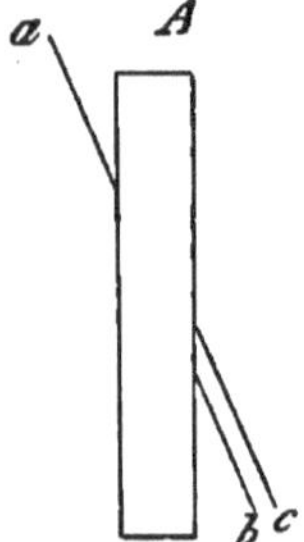

Fig. 115. Continuation of *a* seems to be *c*. Why?

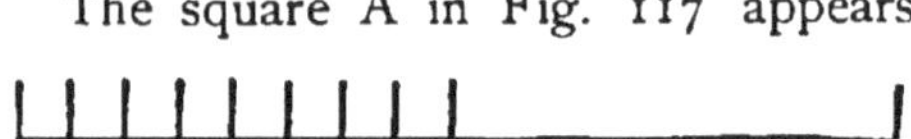

Fig. 116. The Interrupted Distance.

The square A in Fig. 117 appears too long and B appears too tall for the same reason.

Fig. 119 shows the same illusion for angles.

It is evident from these facts why women like to have

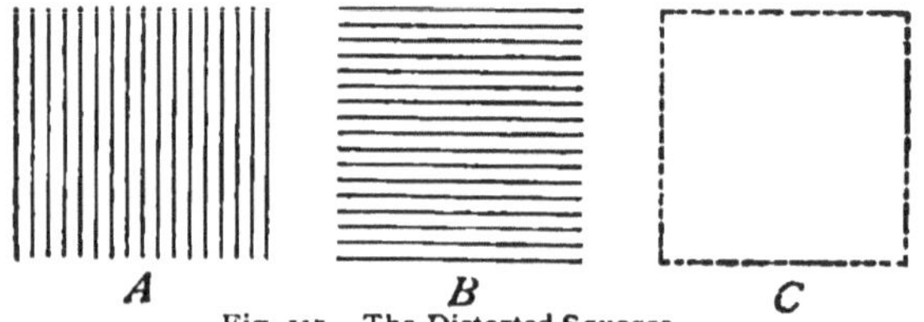

Fig. 117. The Distorted Squares.

Illusion in dress.

as many bows, ribbons, buttons, etc., as possible on the dress. The more the surface of the dress is broken up the taller the person. The illusion is heightened by the diversity of colors employed.

In viewing two lines meeting at an angle, the smaller angle is overestimated as

Estimation of angles.

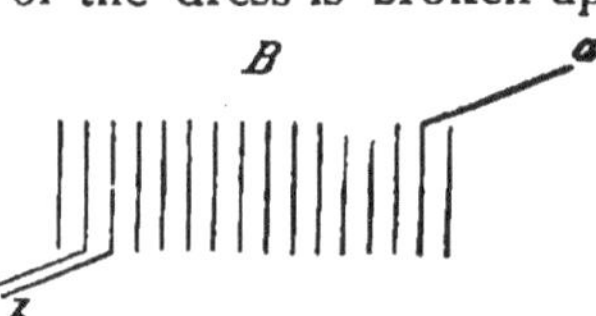

Fig. 118. Which is the continuation of *a*? Why?

compared with the larger. The effect is to press the sides of the smaller angle outward.

Small things and big things.

It is a general law of mental life that small things are

Fig. 119. The Enlarged Angle. Fig. 120. Displacement by Inclined Lines.

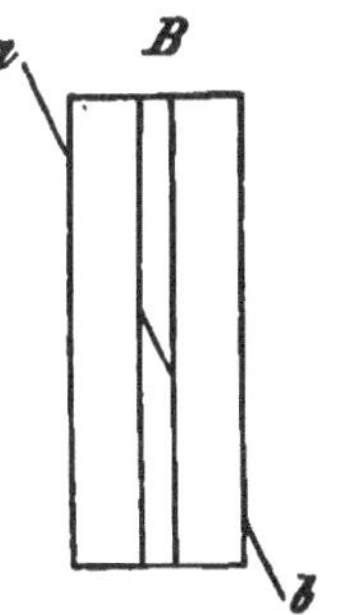

Fig. 121. Why?

thought greater than they are in comparison with large ones. It requires a special effort to realize that a dime is only $\frac{1}{100}$ part of $10.00; one of six pieces of pie seems to be greater than $\frac{1}{6}$ of a whole pie.

Illusions by angles.

The two horizontal lines in Fig. 120 do not seem to be parts of the same straight line because the acute angles are overestimated and the lines are apparently bent from the horizontal. A striking method of showing this illusion is to draw a horizontal line on a

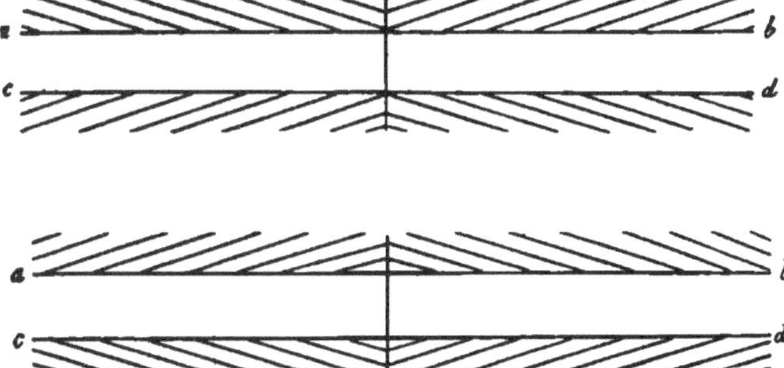

Fig. 122. Breaking Parallel Lines.

slate and then after drawing two inclined lines, as in the figure, to erase the middle portion. In spite of the fact that the two horizontal lines are known to belong to the

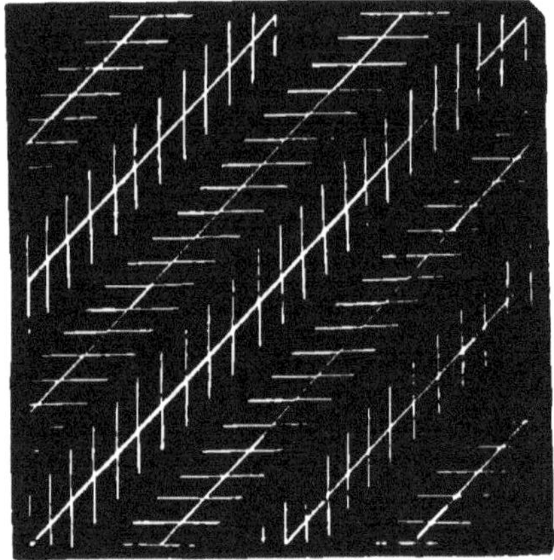

Fig. 123. Tipping Parallel Lines.

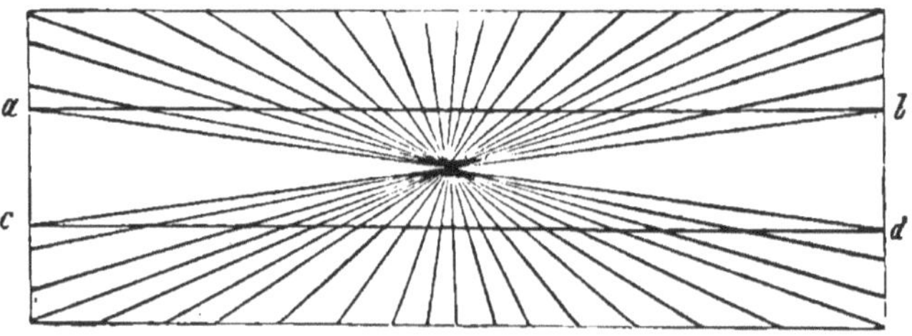

Fig. 124. Bending Straight Lines.

same straight line the illusion is irresistible. This tipping of a line in the direction in which an acute angle points is intensified when a number of angles are made, as in Fig. 122. The top line, for example, has acute angles above it which tip it downward toward the right and

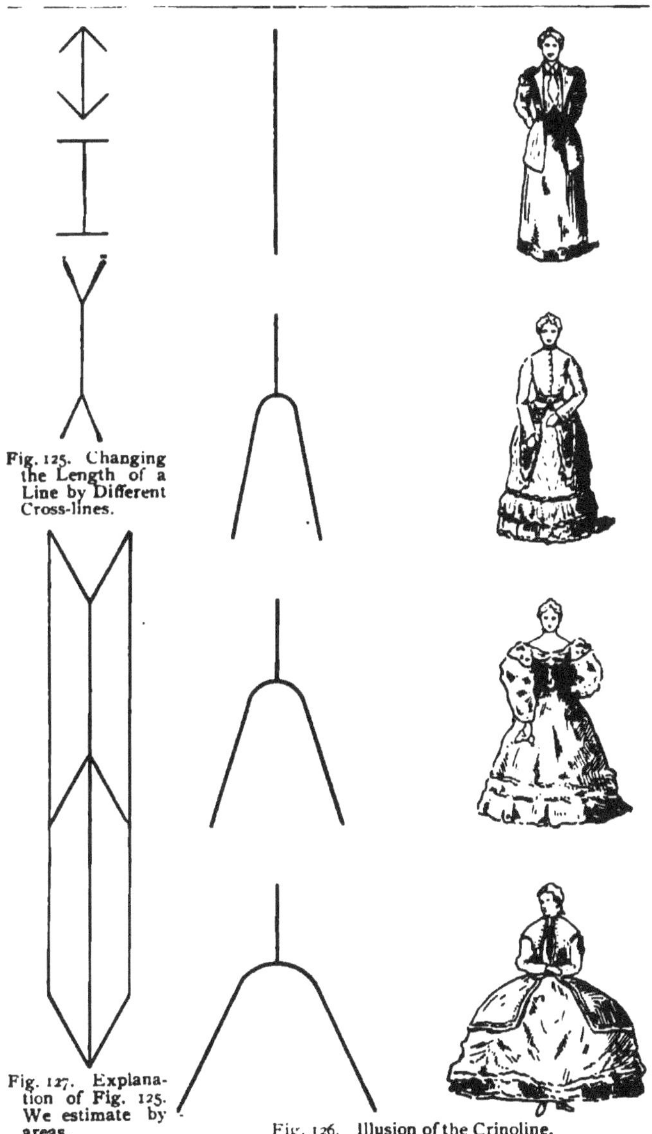

Fig. 125. Changing the Length of a Line by Different Cross-lines.

Fig. 127. Explanation of Fig. 125. We estimate by areas.

Fig. 126. Illusion of the Crinoline.

acute angles below it which tip it upward at the left. The second line is affected in the opposite way. The horizontal lines are really parallel. See also Fig. 123.

Illusion of attention.

A continually increasing change of direction is shown in Fig. 124. The two horizontal lines appear curved.

There is still another class of illusions resting on a mistake of attention. The vertical lines in Fig. 125 are all of the same length, although apparently quite different.

Fig. 126 shows how the crinoline makes people appear shorter.

Explanation.

The reason for this illusion is this: whereas we suppose ourselves to be comparing the vertical lines, we are really paying attention to and comparing the areas between the cross-lines. The areas between the cross-lines in Fig. 127 are about equal, and we judge both parts of the line to be equal, whereas measurement shows them to be unequal.

In Fig. 128 the judgment of height is influenced by the total space occupied by the person's clothing.

Why do tall men dress in sober colors and wear frock coats, while short men prefer dark cutaway coats with silk vest and light trousers? Both kinds of illusion, that of interruption and that of mistaken attention, come into play.

Fig. 128. Why the Bicycle Girl Appears so Short.

Mistaken attention.

Mistaken attention raises the dots in Fig. 129 above their true place at the level of the lower line.

Fig. 129. The Attracted Dots.

The appearance of Fig. 130 depends entirely upon attention.

Illusion of contrast.

Still another source of illusion is contrast between length and breadth. Broad things seem shorter.

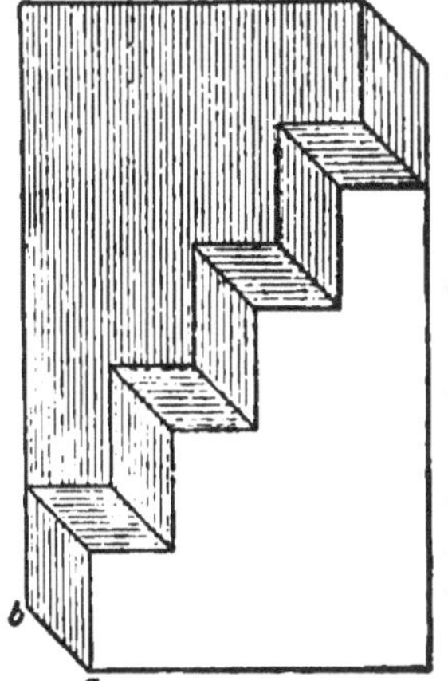

Fig. 130. An Overhanging Cornice, or a Stairway?

Sir Roger's mistake.

"The vast jetting coat and small bonnet, which was the habit in Harry the Seventh's time, is kept on in the yeomen of the Guard; not without good and politic view, because they look a foot taller and a foot and a half broader; besides that, the cap leaves the face expanded and consequently more terrible and fitter to stand at the entrance of palaces." This is Sir Roger de Coverley's observation; how would you explain his mistake? See Fig. 131.

Depth of the world seen with one eye.

Up to this point nothing has been said about the depth or distance of objects. Is the world of one eye a flat surface?

On entering into a strange house with one eye bandaged it is difficult to obtain an accurate idea of the distance of objects. The whole place seems almost flat. Looking out of a window with one eye, the view appears almost as if painted directly on the window-pane.

Size and distance.

We know from experience that objects decrease in size as they recede. From the rear platform of a railway train, the houses, signals, persons, tracks, etc., can actually be seen to shrink together. If we know the actual size of an object we can estimate the distance; if we know the distance we can estimate the size.

Foreground.

In estimating the distance of unknown objects we are guided greatly by the view of the ground in front of them. Thus a tree seen down the road can be roughly estimated in height because the objects along the road afford an indication of the distance.

Illusions of size and distance.

Since our opinion of the size of an object depends on the apparent distance, any illusions of distance will produce illusions of size.

Horizon is far away.

The fact that subdivided distance appears greater than undivided distance was illustrated in Fig. 112. In looking toward the horizon, the glance meets innumerable objects that break up the space, whereas in looking directly upward we find a perfectly clear space. Consequently objects in a horizontal direction ap-

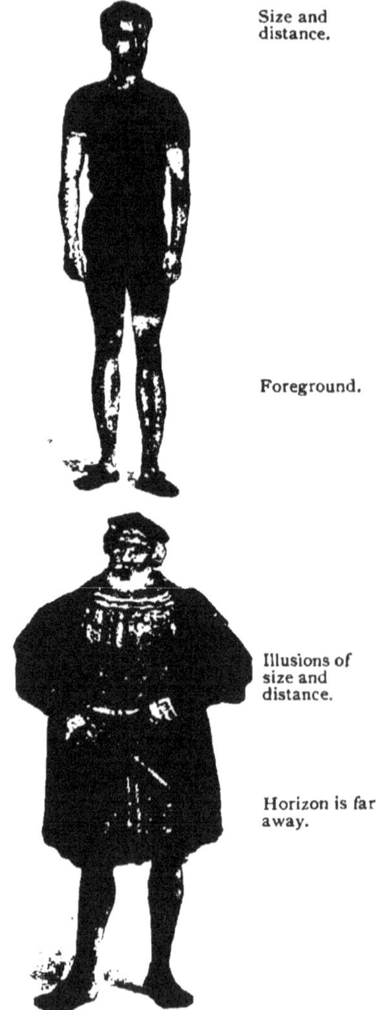

Fig. 131. Illusion of the Yeomen of the Guard. The men are of equal height.

pear more distant than objects in a vertical direction.

Shape of the sky.

For this reason the sky does not appear like the inner surface of a ball, but like the under side of a watch-glass. The amount of this flattening is readily determined. Stars lying 23° above the horizon are apparently half way toward the top. In Fig. 132 you are standing at A. Lines are drawn from A at an angle of 23° to the flat

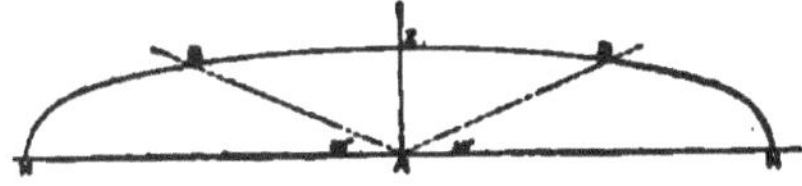

Fig. 132. Shape of the Sky.

ground HH on which you stand. The sky must therefore appear of such a shape that a line drawn from B to H is half the distance from H to Z. Such a surface is indicated by the curved line.

Why the moon changes its size.

Owing to the objects seen on the earth, the moon ap-

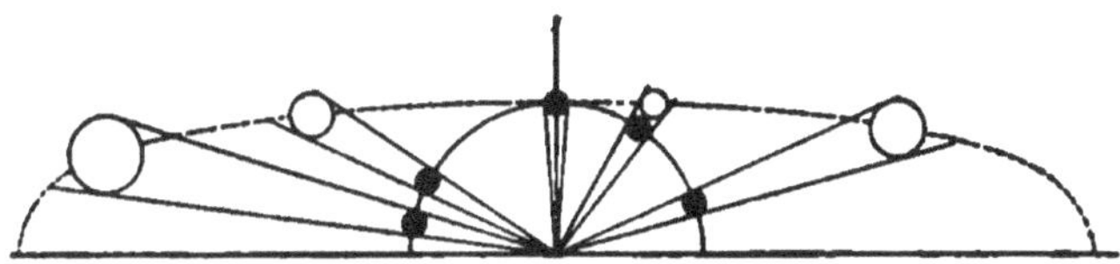

Fig. 133. The Moon Illusion.

pears to be much further away when it rises than it does when it is overhead with nothing between. The moon is seen by the eye as the same in both cases, but the moon near the horizon is apparently larger because it seems further away.

Shades and shadows.

Another means of judging distance is found in shades and shadows. With one eye closed and with the back to the light, hold a mask, preferably painted inside, so that the seeing eye looks directly into the inside. If no shadows are cast, the eye is unable to tell whether it is looking at the inside or the outside. For example, the nose will at one moment appear to be a hollow nose pointing away from the observer and at the next a solid nose pointing toward him. But the moment a shadow is allowed to fall by a change of light, the eye knows at once that the hollow side is turned toward it.

Unclearness of the air.

Another influence regulating our estimate of distance and therefore of size is the unclearness of the air. The air nearly always contains a quantity of mist which makes objects bluer and more indefinite as the distance increases.

In the perfectly clear air, such as is common in the dry regions of the Rocky Mountains or in portions of Maine and Canada, the distance of objects is often quite a puzzle. A canoeist on a lake in such an atmosphere cannot tell whether an island in front of him is one mile or ten miles away.

When a dweller from a dry or moderate region visits the sea-coast, he is subject to great deceptions. The Hudson River at Tappan Zee is wider than at Twenty-third Street in New York City ; yet the latter distance usually appears the greater, owing to the haziness of the coast atmosphere. The illusion disappears, of course, on a clear winter's day.

The coast-dweller is subject to the opposite illusion in the mountains, and innumerable tales are told of travelers who start for a before-breakfast walk to a neighboring hill which is really twenty miles away.

Those who have come into a cloud while ascending a mountain will remember that a small wood-pile looks like a barn, a cow appears larger than an elephant, men are giants, etc. Painters use "atmosphere" to show the distance of objects in a landscape.

Association.

There are also illusions of both size and distance due

to association. Clocks and flags on towers appear much smaller than they really are, because we are accustomed to house clocks and moderately sized flags. The clock of the Battell Chapel as seen from the Yale campus at a distance of 200 feet appears about two feet in diameter ; its actual size is ten feet.

Shadows. A tall object casts a longer shadow than a shorter one. During the greater part of the day the shadows cast by the sun are of moderate size, but early in the morning or late in the evening they become enormously large. This exaggeration we cannot resist, and so at those times trees and houses appear much taller than usual.

Emotion. There is another influence to which I think no one has ever called attention, namely, the emotion produced by the object. In dim light, as at night, most persons feel an indefinite uneasiness, which in nervous persons and children often actually amounts to fear and terror. This uneasiness or fear exaggerates the size of the object. On a dark night the mountains around an inclosed lake, *e. g.*, the Lake of Como, assume an overwhelming aspect and appear far higher than by broad daylight or in pleasant moonlight. In approaching a wharf the buildings and posts are imposing in size. The stories of frightened children are not exaggerations, but true comparisons of the apparent sizes of terrifying and non-terrifying objects. A similar reason may explain the "snake stories."

CHAPTER XV.

SEEING WITH TWO EYES.

The world seen in three different ways.

WHEN the eyes in succession are opened and closed rapidly, objects seem to form different pictures for the two eyes. When both eyes are opened, a third view is obtained. The world as seen with the left eye differs from the world as seen with the right eye ; the world as seen with both eyes is again a different matter.

In our usual experience we see the world as a single world, although we have two eyes that see differently. When we lose control over our vision, as in a state of intoxication, the two eyes are liable to act independently and things are seen double.

Two different one-eye views.

The view with the right eye is what would be seen with the left eye if it were moved a short distance to the right, and likewise the left eye sees what the right eye would see if moved toward the left. The pictures differ only in the point of view.

Fig. 134.

Fig. 135.

The view with both eyes has a relief, a rotundity, that is wholly lacking in the one-eye views.

In looking at a book with the right eye we get the flat view as in Fig. 134 ; with the left eye we get the flat view as in Fig. 135. But with both eyes the book appears in relief. We imagine we see the book as in Fig. 136. What we really see is shown in Fig. 137.

The fundamental fact of binocular vision.

This union of two different flat views into a single solid view is the fundamental fact of two-eyed seeing, or binocular vision. The union is unconsciously performed and is irresistible. Why? Let us trace the process step by step.

Uniting images from the two eyes.

Holding the head directly above these two dots, let

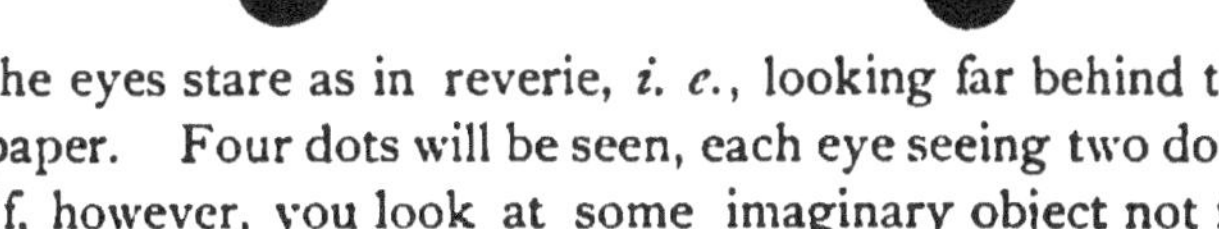

the eyes stare as in reverie, *i. e.*, looking far behind the paper. Four dots will be seen, each eye seeing two dots. If, however, you look at some imaginary object not far behind the paper, the two middle dots will come together. There will then be three dots, the middle one being a combination of one dot from each eye. This can be very plainly seen by sticking the two dots on a window-pane or a piece of glass; when you look at some object at a proper distance beyond the glass, the two middle dots fuse together.

Fig. 136.

Fig. 137.

Eliminating the extra images.

After the union of the two middle pictures into one the two outer ones are still faintly seen. To be rid of these outside pictures all that is needed is to place a strip of paper from the nose to the middle point between the two dots. This makes it evident that the single dot seen is a compound of the dot from the right eye with the dot from the left.

Exactly the same fact is illustrated in Fig. 138, where the problem is to put the bird in the cage. A visiting card is placed from the line AB to the nose, the eyes are relaxed and the bird goes into the cage without difficulty.

The stereoscope.

Most persons find it tiresome or difficult to observe views in the way just described. The presentation of

A

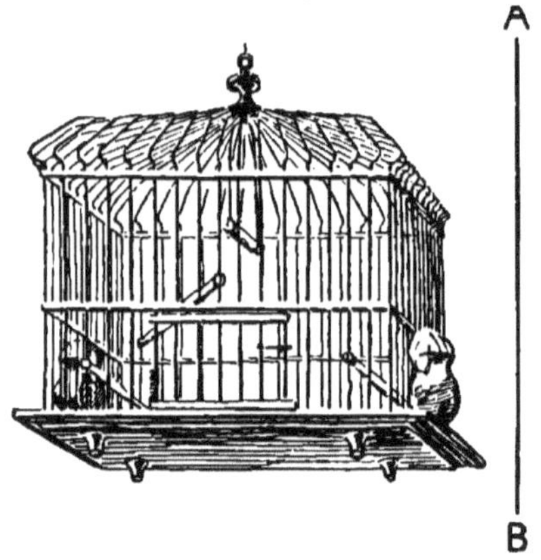

B

Fig. 138. Put the Bird in the Cage by Binocular Vision.

pictures to the eyes separately is most conveniently done by the stereoscope, of which one kind is shown in Fig. 139. A card containing the two pictures is placed on the bottom. The left eye sees only the left-hand picture, the right eye only the right.

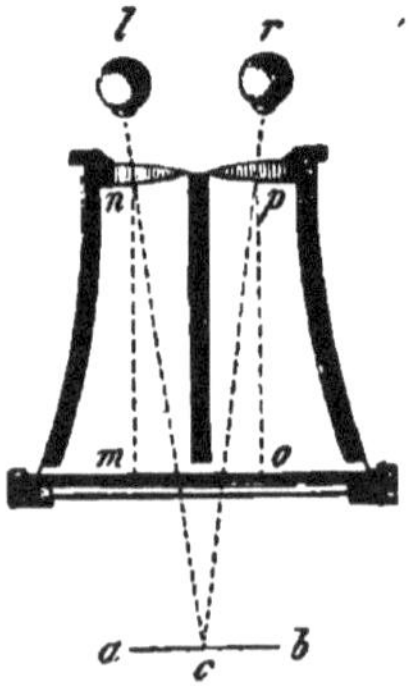

Its principle.

The principle of the stereoscope consists in bringing together the middle pictures for each eye and in avoiding the outer ones. This is most commonly done by means of prismatic lenses.

Fig. 139. The Prism Stereoscope.

Action of the prism stereoscope.

The prism stereoscope contains two glass prisms *n*, *p*, with a partition between and in front of them. It is a property of prisms that an object which is at *m* when directly viewed, apparently changes its position to some such place as *c* when seen through the prism. The amount and direction of the change depend on the character of the prism. Two prisms can be so chosen that for the left eye a picture at *m* is transferred to *c* and for the right eye a picture at *o* is transferred to the same place. The two impressions from different eyes will then be united. The prisms are usually so adjusted that the distance from *m* to *o* is 2½ inches.

The prisms are also lenses.

It is desirable that the prisms should at the same time be lenses, for the following reason. In experimenting with the two dots it will have been noticed that when the gaze was directed to a point beyond them they were seen blurred around the edges. There are very few people who can make each eye look straight forward and yet see near objects distinctly. When looking at distant objects their eyes are far-sighted for near objects. As it is necessary to have the stereoscopic pictures near at hand and yet have the lines of regard parallel, the far-sightedness is corrected by lenses. The two prisms must thus also be magnifying lenses.

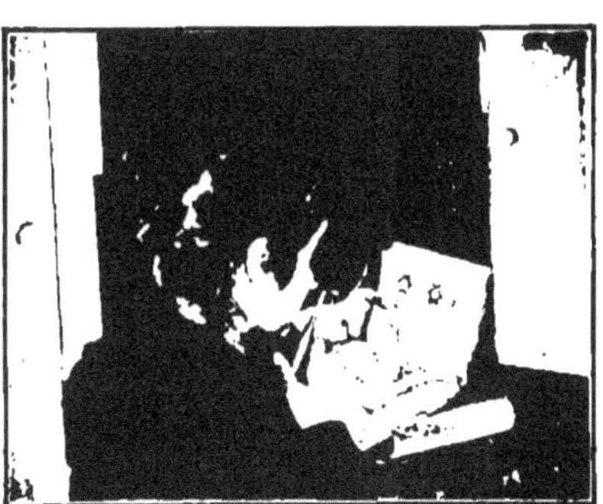

Fig. 140. The Book Stereoscope; How to Use it.

The book stereoscope.

In order to present stereoscopic views to my readers I have had them printed (for the first time) ready for the application of the stereoscope directly to the book. The simplest method is to unscrew the back portion of any stereoscope and hold it to the eyes directly before the picture in the book, as shown in Fig. 140. Three views will be seen (Fig. 141); the one in the middle, C, is the view produced by the combination of views from the right and left pictures, and the others, A, B, are extra views. Another method is to cut off the end of the stick of the stereoscope till the book, when placed against the end, is at just the proper distance. There are no extra pictures in this case.

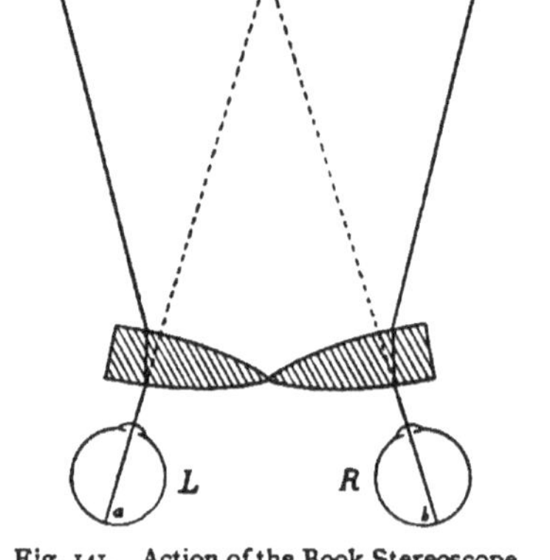

Fig. 141. Action of the Book Stereoscope.

Combining two like pictures.

When two like pictures are placed so that the prisms cause them to fall exactly on the same spot, the images are seen as one. The two heads in Fig. 142 appear as one head.

Combining unlike pictures.

When the two pictures are not alike, they make a compound figure, as in Figs. 143, 144. When two pictures are farther apart than the distance of the middle points of the prisms, they fall beside each other. In Fig. 145 the vertical bars are at the proper distance for union, whereas

Fig. 142. Two Like Pictures.

the horizontal projections are too far apart. The result is a cross. The outline of the vertical bar is darker because the black line of one picture falls on the black line of

Fig. 143. Unlike Pictures to be Combined.

the other, whereas the black line of the horizontal bar in one picture falls on the white space of the other.

Up to this point the results of two-eyed vision have

Fig. 144. Prometheus.

been flat pictures. The production of the effect of objects in relief is not quite so simple.

Double images; crossed disparity.

Let two pencils be held upright before the eyes in a line directly in front of the nose and at about four inches

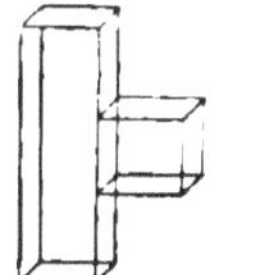

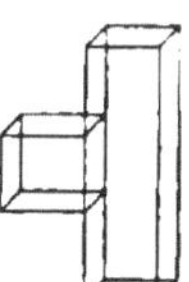

Fig. 145. The Cross.

apart. When looking at the farther pencil you see two nearer pencils, as in Fig. 146. The image L belongs to the left eye because it disappears when that eye is closed; R belongs to the right eye. This condition of the extra images is called crossed disparity; it is to be remembered that objects nearer than the point of regard are seen with crossed disparity.

Fig. 146. Crossed Disparity.

Uncrossed disparity.

On looking at the nearer pencil, the farther appears double (Fig. 147). By closing one eye it is evident that the farther pencil is seen with uncrossed disparity.

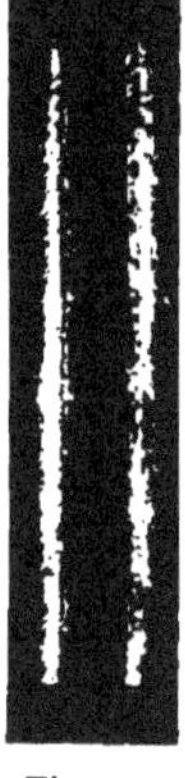

Fig. 147. Uncrossed Disparity.

Relation of disparity to distance.

Thus when we look at any point, the objects nearer than that point are seen with crossed disparity, those farther than it with uncrossed disparity.

Production of relief.

Now hold a single pencil with one end pointing to the nose about two feet away and the other straight in front. Looking at the farther end, you would expect the nearer end to be seen as two ends in crossed disparity (Fig. 148); looking at the nearer end you would expect to see two farther ends in uncrossed disparity (Fig. 149); looking at the middle you would expect to see both ends double in opposite ways (Fig. 150). Since the pencils are continuous to the ends, you would expect the double vision to extend down to the point of regard. What you actually see is one pencil *in relief* (Fig. 151). The continuity of the object transforms the double image into a single one with a new property. By practice it is possible to overcome this union ; a pencil will then be seen as a double one.

Fundamental law of relief. The fundamental law of binocular relief is this : *Two different flat pictures of the same object will be combined*

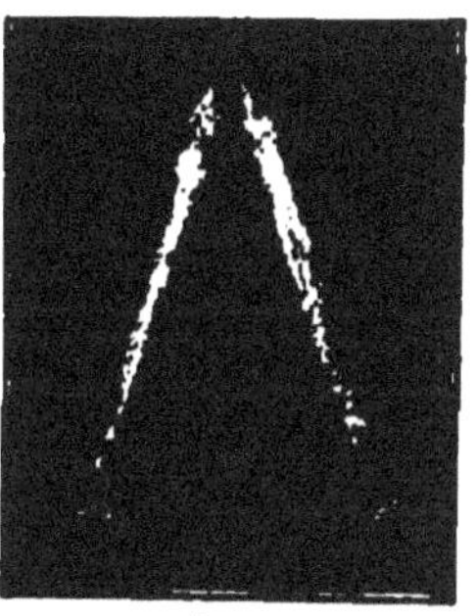

Fig. 148. What we would Expect when Looking at the Farther End.

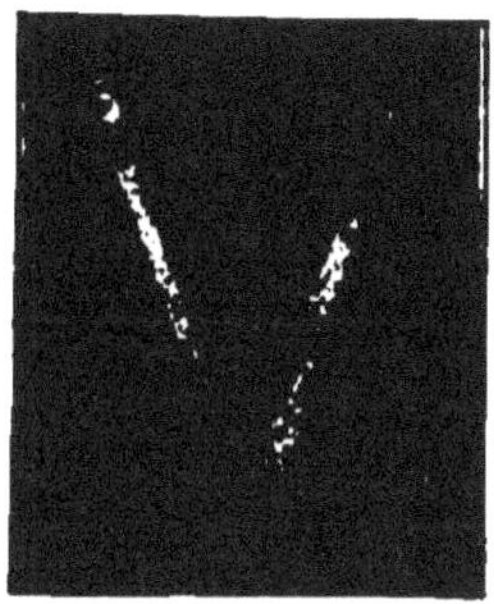

Fig. 149. What we would Expect when Looking at the Nearer End.

Fig. 150. What we would Expect when Looking at the Middle.

Fig. 151. What we Actually See.

into a relief, if each picture is such as would be seen by the corresponding eye singly.

If the two pictures in Fig. 152 are seen with the stereoscope, the result is a union of the two lines into one line slanting away, because the two views are drawn as such a line would appear to the eyes used singly.

If two appropriate views be presented, as in Fig. 153, the result is a figure in relief indicating a pyramidal box.

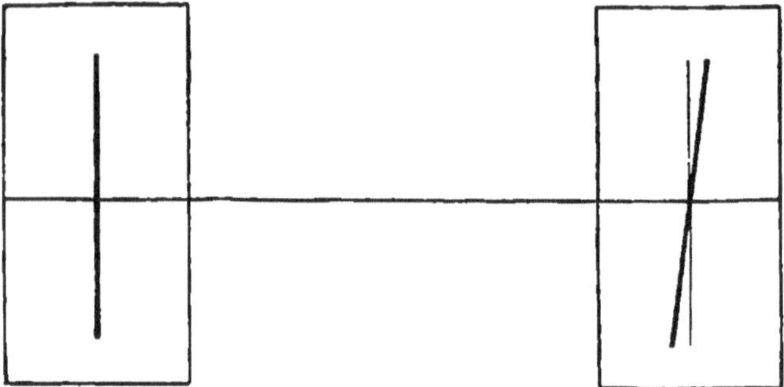

Fig. 152. The Slant Line.

Foretelling the kind of relief by the laws of disparity.

It is possible to tell beforehand whether the box is seen from the inside or from the outside. As the small squares are at the regular distance apart the point of regard is found in the small end of the box. The large

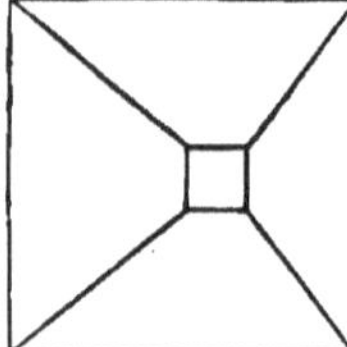

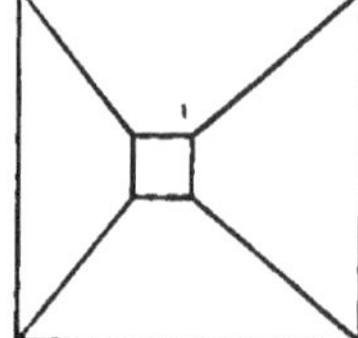

Fig. 153. The Pyramidal Box.

squares are too far apart and are not crossed; this end of the box must be seen in uncrossed disparity. But objects seen in uncrossed disparity are farther away than the point of regard; consequently the large end of the box is farther away. We are therefore looking at the outside of the box.

If the outer squares are drawn so as to be seen in crossed disparity, the larger end of the square is

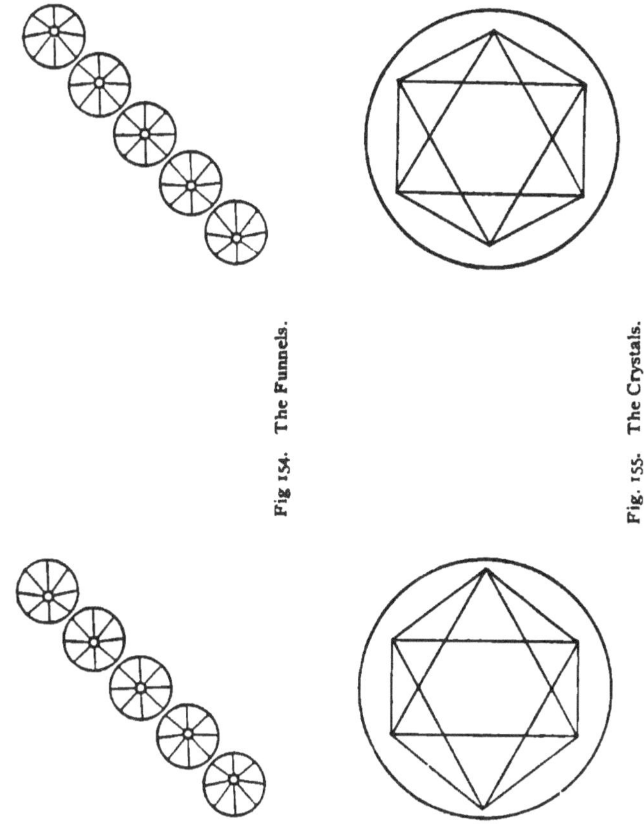

Fig 154. The Funnels.

Fig. 155. The Crystals.

nearer than the smaller end and we are apparently looking at the inside of a box.

Different relations of disparity.

These relations are shown in Fig. 154. The outer circles for each pair are at the proper distance apart and unite to form the base at the point of regard. The smaller circles are seen in different relations of disparity, with the effect that the pictures form a series of funnels, the bottom one being long and pointed toward the observer, the next being shorter but likewise pointed, the

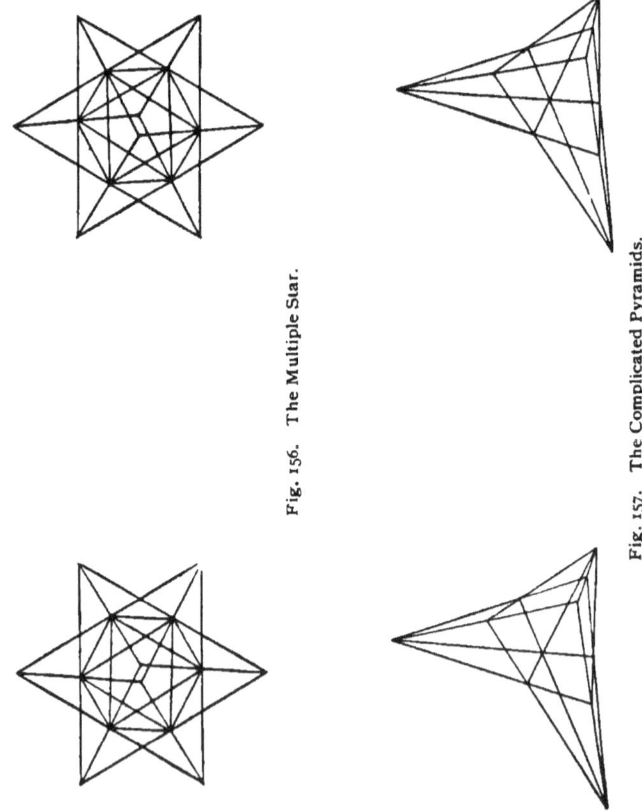

Fig. 156. The Multiple Star.

Fig. 157. The Complicated Pyramids.

middle one being a flat disk, the fourth being short and pointed away, and the topmost one being long and likewise pointed.

From these principles it will be easy to explain the

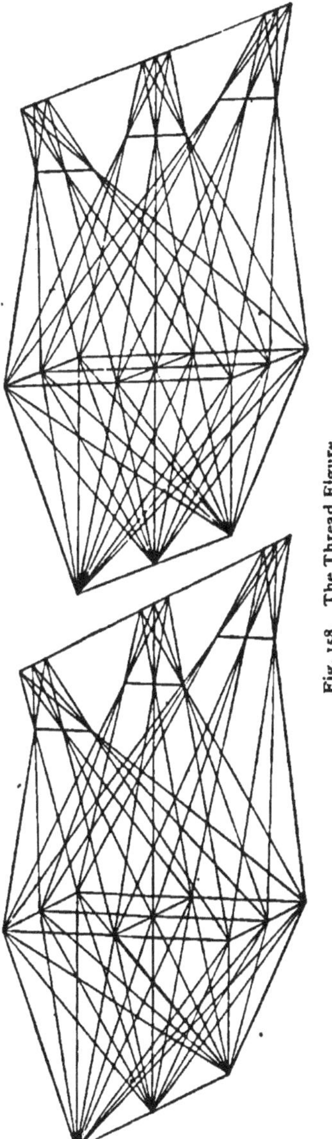

Fig. 158. The Thread Figure.

crystals shown in Figs. 155, 156, 157. The diagram in Fig. 158 is built upon the same laws.

Finally, the stereoscopic views of buildings, persons, and landscapes, such as can be obtained everywhere, are resolvable into the same principles aided by the shading, shadows, and perspective.

Other results of binocular vision.

In addition to the effect of relief which we gain by stereoscopic vision there are several other important results of two-eyed seeing. Among them are: (1) binocular strife, (2) binocular luster, and (3) binocular contrast.

Binocular strife.

Binocular strife is produced when the two different views are separately presented to the two eyes. In Fig. 159 the various rings are filled with lines in different directions; what happens when they are combined with the stereoscope? One of the rings is filled with shading which slants in one direction for the left eye and in the other direction for the right eye. The result is peculiar.

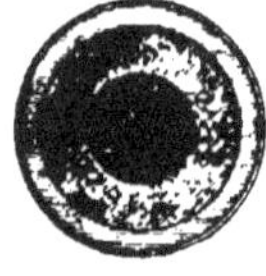

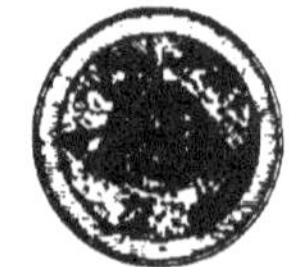

Fig. 159. Binocular Strife.

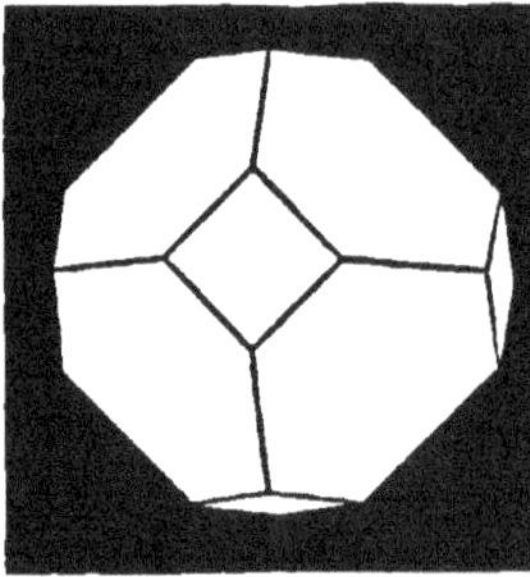

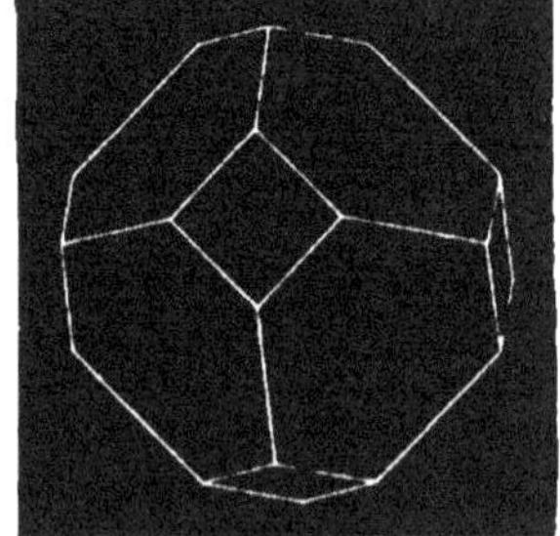

Fig. 160. Binocular Luster. The Translucent Crystal.

Very rarely do the two sets of lines combine to form crossed shading. Sometimes the left-hand shading alone appears, sometimes the right-hand shading wins; generally the two alternate frequently and irregularly. If you happen to think of one kind of shading, that appears. But you cannot keep either kind for more than an instant; the other will replace it. It seems to be largely a matter of attention. Yet, the most frequent aspect of all is that the shading is in patches; the left-hand picture predominates in parts while the right-hand one occupies the rest. And the queer thing about it is that these parts are continually changing. The inner circle behaves in the same way. It is in truth a strife between the two eyes.

Ordinary luster.

Binocular luster, or polish, is so called from the resemblance of the effect to actual polish. A polished object contains a contradiction in itself. Its little marks, irregularities, and corners remain the same, although changes in the position of the light and in the objects near it are followed by changes in the reflection. A polished doorknob differs from an unpolished one by partially reflecting the lights from surrounding objects; there is a strife between the color and general appearance the knob would have if unpolished and the appearance of effects of surrounding objects.

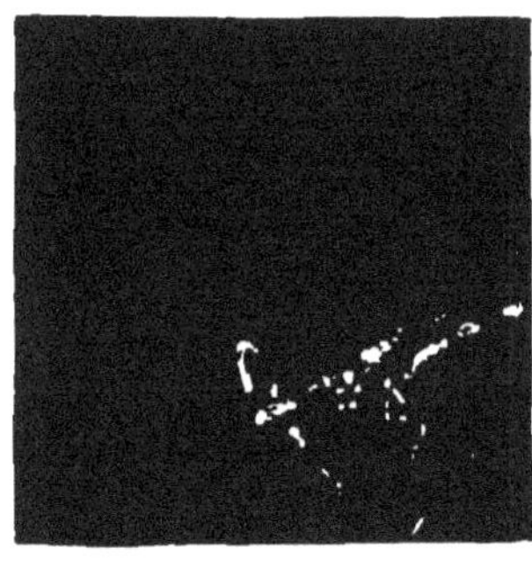

Fig. 161. A Binocular Illustration to Milton's Paradise Lost, Book VI.

Binocular luster.

In Fig. 160 the left eye receives an impression of a white crystal and the right eye one of a black crystal; when viewed with a stereoscope, the same space is covered by a different color for the two eyes. The result is a beautiful, lustrous, translucent crystal, showing changes of light and dark as the binocular strife enters into effect.

Binocular contrast.

Binocular contrast is so called because the result of a binocular strife depends somewhat on the surroundings.

In Fig. 161 we would expect an effect of binocular luster and binocular strife. We do get them, but, in the neighborhood of the most prominent points of each figure, the corresponding color overpowers the other. Thus, in the neighborhood of the angel Michael the white is strongest, while around Lucifer the black overpowers the light.

CHAPTER XVI.

FEELING.

The term, "feeling."

THE word feeling is employed in many meanings. We speak of feeling hunger and thirst, and of feeling pain. We also say that love and hate, joy and sorrow, care and hope, are feelings. We tell of feelings of the beautiful and the ugly, of feelings of truth, honor, and virtue. What is the common property that brings all these into relation? There is one connecting link among them; they express like or dislike.

Like and dislike.

The mental fact which we express by liking or disliking is what we shall term "feeling." It is true that we sometimes say a thing feels hot, feels rough, etc., but we need not fear any confusion with feelings of liking and disliking.

We have thus two simple feelings, liking and disliking. There is no objection to calling them two "qualities of feeling," just as the many rainbow colors can be called "qualities of color," but there is no necessity for doing so.

Indifferent experiences.

Some of our experiences arouse no feeling; they are indifferent. We do not care whether our neighbor wears a fresh-looking coat or a rusty one. The people of Chicago do not care whether their streets show a clear stone pavement or reek with mud. Most sensations, however, arouse some feeling; there are very few things for which we have neither a liking nor a dislike.

The state of our feelings depends on the strength

of the impression that arouses them. For example, a moderately sweet taste, as of sugar, is agreeable ; an intensely sweet taste, as of saccharine, is disagreeable. A moderate degree of saltiness is pleasant, but a strong degree is distasteful. Even a faint bitterness, as in beer, is liked by some persons, while the intense bitterness of quinine is revolting.

Dependence of feeling on the strength of the impression.

Feelings are connected with all sorts of experiences. Muscular exertion, or action of any kind, may arouse feelings. Moderate activity is generally pleasurable ; but tiredness, over-exertion, and unhealthiness, may bring about intense unpleasantness.

Feelings connected with muscular activity.

The extreme pleasure of muscular exercise can be felt only by persons who, like the children in many schools, are forced to remain in one seat for hours. In some class-rooms during a whole morning the children are not allowed to leave their seats ; I have been a pupil in classes where positive terror kept us from making any unnecessary movement. Oh, the joy of jumping down whole flights of stairs after school was over !

From nearly every organ in the body we receive some sensation. The stomach makes itself known by hunger or repletion ; the throat is heard from when thirsty. Each of these sensations may arouse feelings. Thus, hunger and thirst are disagreeable ; repletion and quenching of thirst are agreeable. Other sensations, such as of the liver, were originally very strong, but with advancing culture and age they have to a large extent disappeared. The feelings, however, still remain strong. An overloaded stomach or a disordered liver is liable to make us look upon the world in a very dismal light; the disagreeable feeling from such a source has overpowered all the others.

Feelings connected with hunger and thirst.

Æsthetics of taste.

There are some persons, known as "gourmets," who devote themselves to a study of pleasing combinations of tastes and smells. The fine feeling of the French in this matter has led to the development of the race of French cooks. The puritanical austerity of New England has brought about an almost total decay of the feeling of the beautiful, which exhibits itself not only in its ugly wooden houses and hideously somber garb, but also in its unæsthetical pies, doughnuts, and baked beans.

Influence of touch and temperature on taste feelings.

The influence of touch and temperature on our likings for tastes is so entirely overlooked that scientists have been deceived into supposing that there was some actual chemical difference corresponding to the difference in agreeableness of taste between things which were really mixed with various touch and temperature sensations.

A draught direct from the old oaken bucket has a taste quite different from the same water drunk from a glass. Water from a tin cup is intolerable, yet coffee from a tin cup is far superior to coffee in any other way. The reason is a purely psychological one ; the different sensations of touch and temperature mingle with the sensations of taste to produce agreeable combinations.

Feelings connected with colors.

Various objects are liked or disliked according to their characters. Strong bright colors are always liked. Any one looking at the rainbow colors would be tempted to exclaim, "All colors are beautiful !" This effect is very pronounced when the eye looks directly at the light thrown back by a spectrum-grating (page 160) ; all the colors from red to violet and purple are of an indescribable beauty.

White itself, when not too strong, is just as beautiful. Since we cannot look directly at the sun, the light must be weakened by reflection. This is done by the method described on page 160. White, as seen from such a surface, possesses a beauty as great, if not greater, than the rainbow colors.

Feelings connected with impure colors.

When the colors are mixed with white, less beautiful colors are obtained. No pink can be produced that is equal to pure red; no pale green that is as beautiful as pure green. The whitish skies of our colder climates cannot be compared with the deep blue sky of Italy. When a color or white is darkened, *i. e.*, made less strong, its beauty is lessened. Grays and shades are not comparable with full colors. It is when both these changes are made that indifferent or disagreeable colors are obtained. Grayish pinks or grayish browns or drab blues are somber and unpleasant.

Colored tablets are sometimes given to children with the command to pick out the prettiest one. They generally pick out the yellow, not because (as the teacher supposes) it has anything to do with sunlight, but simply because it is the brightest color in the particular set. With some sets of "spectrum" tablets inflicted on American school children the dull gray red is such a disagreeable color that the children persistently avoid it until the teacher succeeds in producing the desired deformity in the color feelings.

Summary of laws of color feelings.

In general we can say: pure white sunlight, when not too strong, is beautiful; the rainbow colors are beautiful; these all become less pleasing when less strong; the colors become less pleasing when mixed with white; the most disagreeable effects are produced by mixtures of gray (weaker white) with shades (weaker colors).

Among all the good things of life nature is the most beautiful, art is second, and science—why should not science be third? The most beautiful colors and combinations we see are the colors of the spectrum series—science's colors. They lack form; nature makes flower-forms out of science's colors and we have all the glories of the fields; the flowers are nature's colors. Art takes nature's flowers and puts poetry's meaning into them. Flowers as symbols of life, light, and love are art's colors.

Feelings connected with color combinations.

We have thus far spoken only of single colors. When colors are combined, the combination may produce an agreeable or a disagreeable effect, depending on the relation of the two colors.

Combinations of bright colors always agreeable.

In the first place, any combination of the rainbow colors is agreeable. In the rainbow or the spectrum they are all there together. In fact, when colors approach the brilliancy of the rainbow colors, as in stained glass, almost any combination appears fairly good. This is one reason why the patterns in a kaleidoscope have been of so little value in decorative art; for when the colors are most carefully imitated in coarser materials they are apt to lose their brilliancy and to produce disagreeable effects. To a lesser degree this applies also to silk; many color combinations worked out in this material are tolerable on account of their brightness, while the same designs if made in wool or cotton appear very poor.

Most pleasing combinations.

Nevertheless, even with the brightest spectrum colors, some pairs are more pleasing than others. If the colors of the spectrum be arranged in a circle so that complementary colors (page 167) are opposite each other, it can be laid down as a rule established by experiment that a combination of two colors is more agreeable the more nearly they are complementary.

When two grays are combined together, the effect is more pleasing the more they differ. White and black are the most pleasing of all. When a color is combined with gray, or when two colors of different shade or tint are combined, the most pleasing effect is obtained when the difference is greatest. A light red and a dark green will be better than a moderately light red and a moderately dark green. Yet even this last may be better than a light green and a dark blue, because red and green as colors give better effects than green and blue. To get the full effect we should use double contrast: (1) of complementary colors, and (2) of light and dark. For example, we should combine bright red with dark bluish green or dark red with light bluish green, bright orange with dark blue or dark orange with bright blue, etc.

Combinations of grays, and of light and dark colors.

Psychological laws as æsthetical heresies.

It must be confessed that these statements are rank heresies in decorative art. Still they are the combinations preferred by unprejudiced individuals. The bright colors and strong contrasts are preferred by children, by savage tribes, by the peasantry, and also in former periods of art.

Appeal to nature.

Why should we not be allowed to enjoy the combinations of color as nature shows them to us? Nature decorates her fields, hills, and skies with the most gorgeous colors; we northern nations decorate our towns, our homes, and our persons with the dullest combinations we can find. Any one who attempts to put a little life into our colors is decried as an uncultured being. As Ruskin says: "The modern color enthusiasts who insist that all colors must be dull and dirty are just like people who eat slate-pencil and chalk and assure everybody that they are nicer and purer than strawberries and plums. The worst general character that decorative coloring can possibly have is a prevalent tendency to a dirty yellowish green, like that of a decaying heap of vegetables. It is distinctively a sign of a decay of color appreciation."

In these remarks on modern taste I have referred to the tastes of the general public. I must except from

Fig. 162. Single Symmetry, Horizontal. **Fig. 163.** Single Symmetry, Vertical.

them the newer schools of design and also the pretty girls of New York, who have lately taken to the use of harmonious combinations of bright colors.

Feelings connected with form.

The products of art please or displease us not only on account of their color but also on account of their form. The elements of space as exciting pleasure can be

Fig. 164. Double Symmetry.

Fig. 165. Threefold Symmetry.

classed into the division of forms and the direction of bounding lines.

First law of beauty in division of forms.

In regard to the division of forms, we notice first that regular forms are preferable to irregular ones. The simplest kind of regularity is symmetry, *i. e.*, the likeness of the two halves. Horizontal symmetry, *i. e.*, likeness of parts on each side of a vertical line, is the most preferred. Double symmetry is better than single. The more complicated the symmetry becomes, the better we like the result. The degree of symmetry is denoted by the number of lines that can be drawn through the center whereby the half of the figure on one side of the line is just the opposite of the half on the other side. A plain circle is in perfect symmetry in every direction, but it becomes much more pleasing when made into a rosette.

Fig. 166. Fourfold Symmetry.

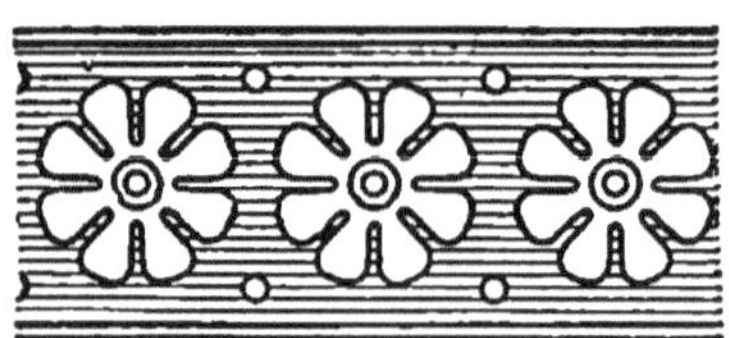

Fig. 167. Eightfold Symmetry.

The pleasing square.

Another kind of regularity is found in a definite relation of height to breadth. The perfect square is very displeasing because, owing to the overestimation of the vertical direction (page 187) the figure appears to be slightly too tall; it seems to impel us to make it correct. As the actual square is shortened we dislike it less, and, finally, when it appears to be a perfect square, we consider it a very pleasing figure.

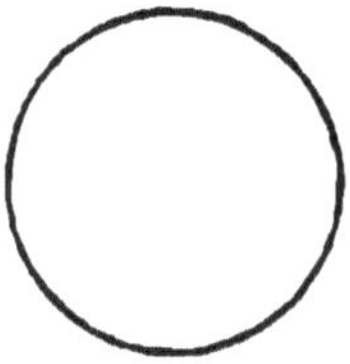

Fig. 168. Perfect, but Simple, Symmetry in All Directions.

Of course, by actual measurement it is no longer a square, but it is a square as far as we are concerned.

The pleasing rectangle.

If a square be changed to a rectangle, it is less pleasing than before, unless there is a certain relation between length and height. Suppose in Fig. 170 the square at X to be successively lengthened in the direction X′. Careful experiments have proven that the degree of pleasure follows some such course as indicated by the line SG. When the relation of the two sides is actually 1 times 1 the figure is very displeasing. When it is equal to an apparent square the pleasure is considerable, S. As it grows in length the pleasure at first decreases, then increases till at a relation 1:1.618 it is at a maximum, G.

Fig. 169. Combinations of Symmetry.

Æsthetics of form.

We have now reached the border-land between psychology and the æsthetics of form, and at the same time we have come to the end of our definite experimental knowledge. The writers on architecture, painting, drawing, and decoration have produced numberless speculations on the psychological principles underlying the beautiful and the ugly. How far each is right we cannot say; as psychologists we have no call to meddle till experimental evidence can be produced.

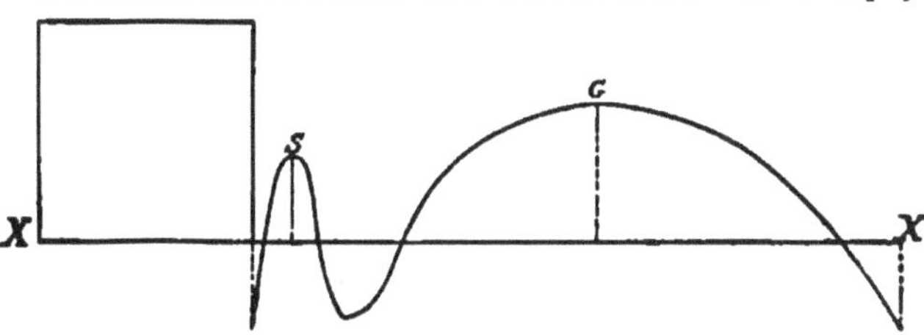

Fig. 170. The Law of Pleasing Relations of the Dimensions of a Rectangle.

"If I were not Alexander, I would be Diogenes." Probably no artist (*i. e.*, form artist or color artist) ever lived who would not choose in the second place to be a musician, a tone artist. What are the laws of feeling in regard to musical tones?

Pleasure in tones.

Perfectly pure tones, like those from good tuning-forks, seem hollow and less agreeable than tones from the musical instruments, which are really compound tones. The tone of a violin, for example, is composed of a main tone and a great number of weaker tones, such as the octave above, etc., whereas the tone of a flute is nearly simple.

Further than this there is very little that can be said in regard to tone-feeling.

Feelings accompanied by internal sensations.

If we look closely at the feelings, especially in their more intense stages, we can hardly doubt that they are invariably accompanied by actual sensations from parts of the body. These sensations often attain an intensity equal to that of the sensations aroused by direct external or internal stimulation. Sometimes they may be even localized with some degree of definiteness. They also show a determinate quality which varies with the general condition of feelings, and which is reflected in the expressions which we employ to describe this. All excessive feeling is attended by physical pain, whether diffused over the body generally or restricted to a particular organ. Moderate excitations also affect the sensations, though less strongly, and are more definitely local-

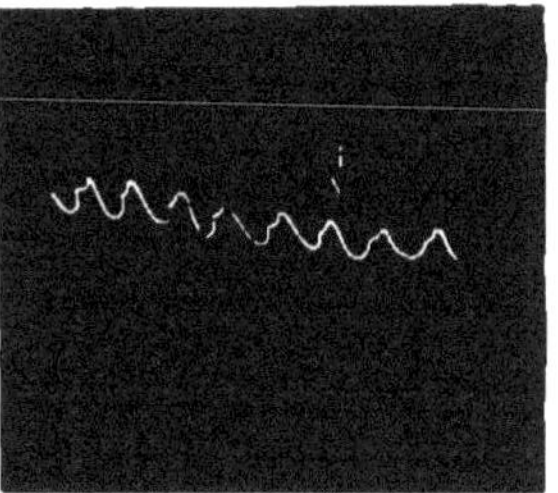

Fig. 171. Change of Pulse as a Result of Pleasure. (The record runs this way : ⟵)

ized. References to this localization of the sensations in particular states of feeling are found in ancient literature. Every passion was supposed to be seated in a particular organ; and it must be admitted that where observation was wanting imagination took its place. Anger was placed in the liver, envy in the spleen, the higher emotions in the organs of the breast. Even today the heart is the seat of the most various affective states. Care and disappointed hope bring on heartache; despair dies of a broken heart; love through all its changes and chances has its source and center in the heart; courage has a lion-heart, and "faint heart never won fair lady."

Poetical localization of the feelings.

The heart's relation to feeling.

There is really good reason for this relation of the heart's activity to the state of feeling; for the heart nerves are those most easily excited by changes in our feelings. Every excitation of feeling manifests itself in a weaker or stronger, quicker or slower, heart-beat. Joy and hope make the pulse quick and strong. The pulse-line, as traced on a smoked drum, rises as the intensity of the beat is stronger, and the beats come more rapidly when the person experiences a very pleasant feeling. Care and anxiety render it weak and slow; terror arrests it altogether. And there are many indications that other organs react to such changes of feeling. It has often been noticed that violent anger results in a return of the bile to the blood, which means a derangement of the

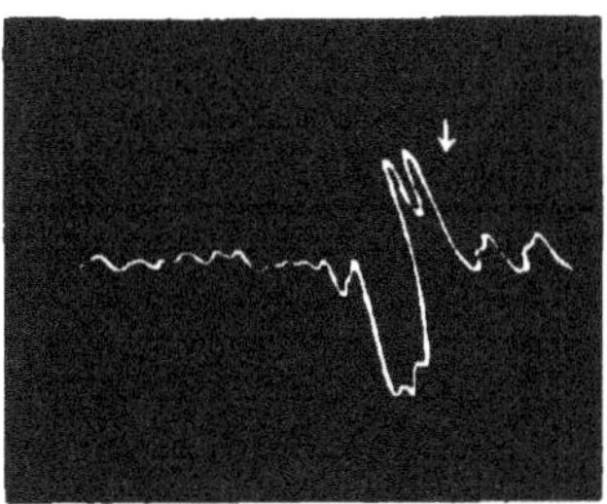

Fig. 172. Change of Pulse as a Result of Anger. (The record runs this way: ←)

Relation of other organs to feeling.

function of the liver. The tear-glands are very easily excited by the feeling of sorrow. And we should undoubtedly discover other similar connections were it not that they have no external symptoms. Besides the particular organ which is especially concerned in a particular state of feeling, there are always other organs more or less affected; and it is the complex of sensations resulting from the sum total of these separate sensations that constitutes the mass of internal feelings and sensations. The muscles, for example, are almost always involved in this secondary excitation. We have direct experience of the energy and tension, or of the exhaustion and relaxation of our muscles; and our general states of feeling are altogether different according as the limbs are movable and elastic or are heavy loads to weigh the body down. A feeling of joy and excitation makes movement easy and prompt; a depressing feeling renders it slow and heavy.

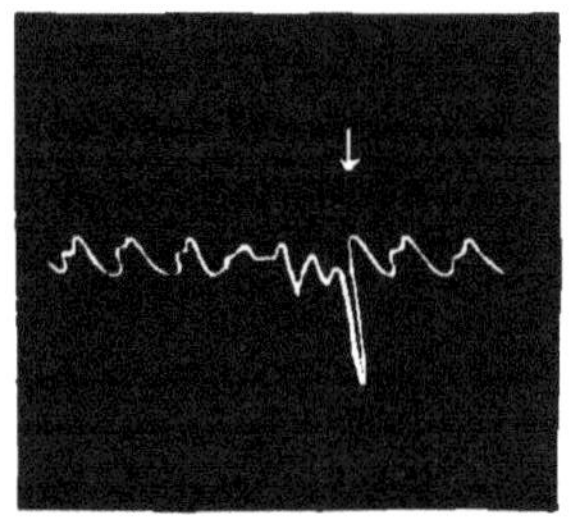

Fig. 173. Change of Pulse as a Result of Fright. (The record runs this way: ⟵)

The muscles.

CHAPTER XVII.

EMOTION.

Feelings influence thought.

THE preceding chapter treated of the simple cases of liking and disliking. But when we like anything, our course of thought generally becomes different from what it would have been if we had disliked it; and likewise the reverse.

Emotions are complex.

The complex processes of thought and feeling combined are generally called emotions. They are among the most important mental phenomena, exerting a marked influence both upon thought and voluntary action. They are reflected in certain expressive movements. These are further connected with reactions of the heart, the blood-vessels, the respiratory muscles, and certain secretory organs, which take on a special characteristic form in each particular emotion.

Their wide influence.

The general subject of the emotions has been so clearly stated by Wundt that, at the present stage of psychological investigation, I cannot do better than follow his treatment for the rest of this chapter.

Analysis of an emotion.

The typical emotion has three stages: an initial feeling; a subsequent change in the train of ideas, intensifying and qualitatively modifying the initial feeling; and (always supposing that the emotion is distinct and well defined) a final feeling, of greater or less duration, which may possibly give rise to a new emotion of which it forms the initial feeling. The principal difference between feeling and emotion consists in the second stage —the alteration in the train of ideas. The presence of this alteration enables us to divide emotions into two classes, excitant and inhibitory. Instances of the former are joy and anger; of the latter, terror and fear. At the same time, all very intensive emotions are inhibitory in character, and it is only when they have run some part of their course that their excitant side comes to consciousness. On the physical side, the effect of emotion upon the train of ideas is accurately reflected in external movement. The excitant emotion quickens thought and involves heightened movement of face and limbs, increase of heart activity, and dilating of the blood-vessels; the inhibitory emotion paralyzes, or at least relaxes, the muscles, slows the heart-beat, and contracts the vessels. All these physiological effects have their accompaniment of sensations, which intensify the feelings in the emotion.

Two classes of emotions.

Physical effects of emotion.

Less intensive degrees of emotion are called moods. It is a general rule that the duration of emotion varies inversely with its intensity; so that moods are more permanent states of mind than emotions proper. Violent emotions are sometimes termed "passions." The name indicates that strongly emotional states, which oscillate between liking and disliking, tend invariably toward the side of the latter. "Passion" also implies that a particular emotion has been habitual. Hence the word is often used to denote a permanent condition which finds its expression in frequent outbursts of emotion.

Moods.

Passions.

The most indefinite emotions are joy and sorrow. When sorrow is directed upon the external object which excites it, we call it care. We can only be careful about others, and if we wish to express the fact that an object arouses

Sorrow.

Care. no interest in us we say that we do not "care" about it. The personal opposite of care is melancholy. The Melancholy. melancholiac is centered in himself; he withdraws from the world to brood in solitude over his own pain. Care and melancholy become anxiety and dejection when they pass from emotions to permanent moods. Intermediate between these objective and personal forms of sorrow stand gloom and depression. We may be gloomy as to our fate in the world and depressed about a loss we have suffered, or we may be gloomy or depressed without any external reason, simply because our mood will have it so.

Fig. 174. Sorrow.

Gloom.

Joy. Joy, like sorrow, assumes different forms according to the direction which it takes. But we have not nearly so many words to express joyous emotion as we have to express sorrow. A joyous mood we call cheerfulness, or, in its higher stages, hilarity. But we cannot tabulate the joyous emotions as objective and personal, as we could their opposites. It may be that our poverty of words points to a distinction in the facts of our emotional life. The joyous emotions appear to be more uniform, less variously colored, than the sorrowful.

The emotions of joy and sorrow, whether their reference is mainly external or to the person himself, are always personal in character; the emotional excitation of our own minds is always the principal thing. A mood, on the other hand, may be objectified by our putting our own feelings into the external objects which excite them. If joy and sorrow are the expressions of an internal harmony and disharmony, these objective emotions are the result of some external harmonious or inharmonious impression. Like and dislike are the most general forms of objective emotion corresponding to joy and sorrow on the personal side. They further imply a movement to or from the object: what we like attracts us; what we dislike repels us. And this movement finds its expression in the various particular forms in which the general emotions occur.

Personal character of joy and sorrow.

Fig. 175. Joy.

The attraction which a pleasing object has for us all we call charm; a thing is "charming" which both pleases and attracts us. The opposite of charm is repulsion, a violent dislike, which makes us turn away from an object in displeasure. Repulsion becomes aversion, and, at a still higher stage, anger, when it is turned

Charm and repulsion.

directly upon the repellent object; it becomes chagrin and mortification if the unpleasant mood can find no outlet. The extreme degree of anger is rage; the extreme of mortification is exasperation. The opposite of chagrin is contentment; when pleasantly concerned with external objects it becomes delight; when quietly occupied with its own affairs, happiness.

Fig. 176. Anger.

Indifference.

The two opposite processes of charm and repulsion find a meeting-point in indifference. Indifference has a tendency in the direction of unpleasantness; when sense or thought is sated with the indifferent or perhaps originally attractive object, it passes over at once into repugnance. Repugnance is as much sensation as emotion. In the latter shape it has an objective form, antipathy, and a personal one, discontentedness. If the emotion becomes a permanent mood, we have weariness and dissatisfaction.

Distinction between sensation and moods, or emotions.

In all these cases, emotion and mood are at once distinguishable from sensation by their connection with a train of strongly emotional ideas. When we feel joy or sorrow, our mood is the result of some pleasant or painful experience which may be resolved into a number of ideas. If we are mourning the death of a friend, our consciousness is filled by affectionate memories, more or less clear or distinct, which coöperate to produce the emotion. If we are made angry by some insolent remark, our first feeling is one of violent displeasure; then our mind is flooded by a torrent of ideas connected with ourselves, the personality of our assailant, and the more immediate circumstances of the insult. Most of them will not attain to any degree of clearness, but all are held together by the feeling of displeasure, which in its turn is intensified by the sensations accompanying our expressive movements.

A simple sensation which has no special relation to our past mental history will, therefore, hardly be able to excite an emotion, though it may call up quite intensive sensations. Where an emotion appears we may assume the presence of memory-ideas, of experiences in which a similar sensation was somehow concerned. The full and harmonious tone of a peal of bells sounds holiday-like to us, because we have been accustomed from childhood to interpret the chimes as harbingers of holidays and religious festivals ; the blare of the trumpet reminds us of war and arms ; the blast of the horn brings up the greenwood and the tumult of the chase ; the chirping of the birds tells us that spring has come ; the chords of the organ suggest a congregation assembled for devotion.

It is probably memory again which determines the way we feel in regard to color impressions, although in their case the ideas aroused are not so clear or distinct. Why is white the color of innocence and festivity, black the color of mourning and severity ? Why do we choose blood-red to express energy and spirit, or purple to express dignity and solemnity ? Why do we call green the color of hope ? It would be difficult to trace the mood to its original source in each particular case. In many cases it probably arises from an obscure association of the color with the occasions when custom prescribes its use. Purple has been the royal color since time began ; and black is almost everywhere among the western nations the color of the mourner's garments.

Original connection not explained by association.

It is true that this association does not fully explain the connection between the sensations and the mood which it arouses. There must be some original reason for the choice of one particular color, and no other, as the expression of a state of feeling. It is perhaps justifiable to look for this reason in the relationship between the sensation and the emotional character of particular colors.

Emotions of the future.

Emotions exhibit peculiar modifications when their character is not determined, as in the cases hitherto considered, by impressions and ideas belonging to the present and considered as present, but by ideas which refer to the future.

Expectation.

The most general of these emotions of the future is expectation. In it we outrun in impressions of the present and anticipate those which the future will bring. We look forward to its realization ; and if this realization is postponed, it becomes what we call strained expectation ; the bodily feeling of strain accompanies the emotion. In expectation the muscles are tense, like those of a runner awaiting the signal for the race, although very possibly the expected impression demands no motor response whatsoever. Expectation becomes watching if the expected event may happen at any moment, and our sensory attention is wide awake to prevent its passing unnoticed. The tension is relaxed with the appearance of the expected impression. If the occurring event fulfils our expectation, we have the emotion of satisfaction ; if not, that of disappointment. Satisfaction and disappointment constitute sudden relaxations of expectant attention. If expectation is prolonged, its tension will gradually disappear of itself, for every emotion weakens with time.

Satisfaction and disappointment.

The opposite of disappointment is surprise. Surprise is the result of an unexpected event. In it we have ideas suddenly aroused by external impressions, and interrupting the current train of thought in a way which we did not anticipate, and which, at the same time, strongly attracts our attention. Surprise may be in quality pleasurable, painful, or altogether indifferent. A special form of it is astonishment. Here the event is not only unexpected at the moment, but unintelligible for some time afterwards. Astonishment is, therefore, a kind of continued surprise. If it passes into a still more permanent mood, it becomes wonder. **Surprise.**

The feeling of rhythm, which is the single psychological motive in dancing and ranks with harmony and disharmony as a psychological motive in musical composition, contains the elements both of expectation and satisfaction. The regular repetition in rhythmical sensations makes us expect every succeeding stimulation, and the expectation is immediately followed by satisfaction. Rhythm therefore never involves strain, or, if it does, it is simply bad rhythm. In pleasant rhythms satisfaction follows expectation as quickly as possible. Every impression arouses the expectation of another, and at the same time satisfies the expectation aroused by its predecessor, whose relations of time it reproduces. Rhythm is an emotion compounded of the emotions of expectation and satisfaction. A broken rhythm is emotionally identical with disappointment. **Feeling of rhythm.**

Hope and fear may be regarded as special forms of expectation. Expectation is indefinite. It may refer to an event desirable or undesirable, or perhaps relatively indifferent. Hope and fear decide expectation ; hope is the expectation of a desirable result, fear the ex- **Hope and fear.**

pectation of something undesirable. It is hardly correct to call hope a future joy, or fear a future sorrow. The feelings can as little penetrate into the future as the senses. Hope and fear are the expectation of future joy and future sorrow, but not joy and sorrow themselves. Either of them may be realized, just as expectation may lead to satisfaction or disappointment.

Alarm and fright.

Fear of some immediate disagreeableness is called alarm. Fright bears the same relation to alarm as does expectation to surprise. Fright is the surprise occasioned by some sudden, terrifying occurrence. It becomes consternation when the occurrence physically paralyzes the individual experiencing it; and it is called terror when he stands amazed before the event. Consternation is, therefore, the more subjective side of fright, and terror its objective side. If fear is continued, it becomes uneasiness. The uneasy mind is always afraid; every occurrence alarms it. In other words, the emotion has become permanent, but at the same time somewhat less intensive.

Fig. 177. Fright.

Intellectual emotion.

The emotions both of the present and future assume the most varied forms, according as the idea changes. Especially important are those attaching to certain intellectual processes and originating in the peculiar feelings which accompany them. We can distinguish four kinds of intellectual feelings: the logical, ethical, religious, and æsthetic. Attaching themselves to very complicated connections of ideas, they almost invariably pass over into emotions, and in that form exert upon our mental life an influence which far exceeds that of any other state of feeling. Their analysis belongs, of course, to the special sciences from which they have their name. We will devote a few words to the logical emotions; first, because they are often overlooked altogether, and, secondly, because their relationship to the emotions of the future enables us to use them as illustrations of the passage of emotion in general into the particular forms of intellectual emotion. Logical emotions are those connected with our current of ordinary thought.

Logical emotions.

It is well known that the rapidity of the course of thought exerts a considerable influence upon our general emotional condition. It is not indifferent to us whether our ideas succeed one another at their normal rate, or proceed slowly with many restraints and interruptions, or pour in upon us in perplexing confusion. Each of these cases may be realized, whether from internal or external causes. Our state of mind at the moment, the topic of our current thought, and external sensations may all be of determining influence. The traveler in a new country is well content when his carriage takes him quickly from one impression to another, not so quickly that he cannot assimilate what he sees, but not so slowly that he is always wishing himself farther on amid new scenes. He is not so satisfied if he is lumbering along in a heavy wagon, passing for days through the same scenery, when he longs to be at his journey's end, or is curiously anticipating novel experiences. Nor is he quite happy when the railway takes him swift as an arrow through a country rich in historical associations, and he tries in vain with deafened ears

Emotional influence of the rapidity of thought.

External causes of change in rapidity.

and tired eyes to fix some of its features in his memory.

Internal causes.

This general result can be produced by internal causes just as well as by the variation of external impressions. If you have to solve a mathematical problem in a short time, your thoughts trip each other up; you are in a hurry to get on, but are obliged to go back, because you have been following out a second thought before you had finished the first. And it is not less disagreeable to be stopped in the middle of your task because your thought halts, and you cannot answer the next question. On the other hand, work becomes a recreation when one result leads certainly and easily to another.

Three emotions of thought.

We have, therefore, the three emotions of confused, restrained, and unimpeded thought. The last two are closely related to the emotions of effort and facility. Correlated with these are the sensations attached to ease and difficulty in muscular action. They are generally present to some degree in the corresponding emotions, even when the causes of these are wholly mental. The feeling of effort is a weight which presses upon the emotional condition; and its removal is accompanied by a sudden feeling of pleasure. This characteristic feeling of relief affects us mainly by way of contrast to our previous mood.

Effort and facility.

Enjoyment and tedium.

Special forms of the emotions of unimpeded and restrained thought are those of enjoyment and tedium.* In enjoyment our time is so well filled by external or internal inducements to activity of ideas that we hardly notice its passage, if we do at all. The nature of tedium is indicated by its name. Our time is unoccupied and passes slowly because we have nothing else to think of.

* This is the dignified word for "boredness."

Tedium, therefore, has a certain affinity to expectation, but it is an expectation that has remained indefinite. It does not expect or anticipate any particular occurrence, but simply waits for new events. A long continued expectation always passes into tedium, and an intensive tedium is hardly distinguishable from strained expectation.

Failure and success.

Related to the feelings of effort and facility are those of failure and success. Investigation and discovery are attended by feelings which show a close resemblance to those of effort and facility. The feelings of agreement and contradiction are somewhat different. They originate in the comparison of simultaneous ideas, which in the one case are accordant, and in the other refuse to be connected.

Fig. 178. Very Doubtful.

Doubt.

Doubt, which we can consider as an oscillatory feeling, is not the same as contradiction. The doubter cannot decide which of two alternatives is the correct one ; he is in contradiction with himself. The conflicting ideas are nothing real, but simply products of his own thought, so that there is always the possibility that the contradiction in doubt may be resolved by experience or more mature consideration; and so far doubt is related to the emotions of the future. This relationship becomes still more apparent in a special form of doubt—the feeling of indecision. When we are undecided we are in contradiction with ourselves as to which of different roads we shall follow, or which of different actions we shall choose. Indecision is therefore a doubt implying reference to action and resolved by it.

CHAPTER XVIII.

MEMORY.

If I were writing a dictionary I would define memory as that portion of mental life about which everybody has been talking for three thousand years without telling us anything more than anybody of common sense knows beforehand.

What is memory?

By memory we mean the relation between two ideas occurring at different times, whereby the second is intended to be like the first. In some schools of design the model is shown for a short time, whereupon the pupils are required to draw from memory. The original impression, sometimes called the sense-perception, was that of the model; the memory-picture is the mental picture from which the drawing is made. The relation between the two pictures is what we call memory.

There are numberless entertaining stories concerning great and peculiar memories, but it is much to be doubted if anything of any value is gained by repeating them. Instead of following the beaten path it will be better to enter at once into the experimental work on the subject. Facts first, theories afterwards.

Methods of investigation.

Memory can be investigated in two ways: by measuring the difference of the repeated idea from the original, or by counting the number of successfully repeated ideas out of the total number.

Memory for actions is a good subject to begin with.

Memory for movement.

How accurately does the arm remember a straight movement? With the eyes closed draw on the first sheet of a pad of paper a vertical line of any agreeable length. Without opening the eyes tear off this sheet; it is very convenient to have the pad fixed firmly to the table. After waiting five seconds (if you have no ticking clock at hand, some one can tell you the time), with the eyes still closed draw a second line which you judge equal to the first. Tear off the sheet as before. After waiting five seconds again, draw a third line of the same length as the *second* (you need not attempt to recall the first). Continue in this way till eleven lines have been drawn from memory.

Experiment.

Computing the results.

With a millimeter-scale (or a ruler divided into sixty-fourths of an inch) measure each line. The difference between each line and its predecessor gives the amount of error in remembering after the particular five seconds. Thus, with a line about 100 millimeters long, we might get a series of errors of -2, -1, $+1$, -1, $+2$, -1, -3, -2, -3, -1, where $+$ indicates that the second line was too long and $-$ that it was too short.

In memory there are two changes that go on: first, an actual change in the idea remembered; and, second, an increasing uncertainty.

Average change.

If we average up all the errors, taking into account the signs, we shall get the average change. Thus, the average of the set we have just noticed is

$$\frac{-2-1+1-1+2-1-3-2-3-1}{10} = \frac{-11}{10}$$

$$= -1\tfrac{1}{10} \text{ or } -1.1^{\text{mm}}.$$

This is the average change introduced by the lapse of five seconds.

What is the uncertainty of our judgment? This we

find by averaging all the separate errors without regard to sign; thus

Average uncertainty.

$$\frac{2+1+1+1+2+1+3+2+3+1}{10}=\frac{17}{10}$$
$$= 1\tfrac{7}{10} \text{ or } 1.7^{mm}.$$

We would thus say that the average uncertainty introduced by a lapse of five seconds is 1.7^{mm}.

Dependence on the elapsed interval.

By repeating the experiments with an interval of ten seconds, we find the average memory-change and the average uncertainty due to that interval. Likewise we can use intervals of fifteen seconds, thirty seconds, one minute, five minutes, etc.

Fundamental law.

Simple as such experiments on memory are, there seem to have been only two sets of them, neither of them of any definite value. As the matter is of fundamental importance in the study of memory I have had a set of them made especially for my readers. From the results the fundamental law of memory can be deduced as follows: The average change is an individual matter depending on circumstances, but the average uncertainty increases in a definite relation to the time.

Fig. 179. A Leaf from Daisy's Copy-book.

In learning to write by means of a copy-book the eye gets the mental image and then, looking down, guides the pen. As the distance from the copy to the line grows larger, the eye has time to partially forget the exact form of the lines in the copy (Fig. 179).

The memory for the force of action can be investigated

Memory for power.

with the dynamometer, described on page 79. The pull is executed to any agreeable weight, say ten ounces. After five seconds it is repeated to apparently the same weight. The amount of the error is recorded by some other person. Again after five seconds the pull is repeated, and so on. The average change and the average uncertainty are calculated as before.

Then ten seconds, fifteen seconds, and so on, are used as intervals. We finally obtain the law of memory for force. Here, also, there have been no facts to proceed upon. The results of an investigation lately made show a rapid increase both of the average change and the average uncertainty.

Cross-memory.

The very curious fact of cross-education has been noticed on pages 75, 83, and 112; there is also a "cross-memory."

If the original line in the experiments on page 240 be drawn with, say, the left hand, it can be remembered with the right hand. If the original pull on the dynamometer be made with one hand, it can be remembered with the other.

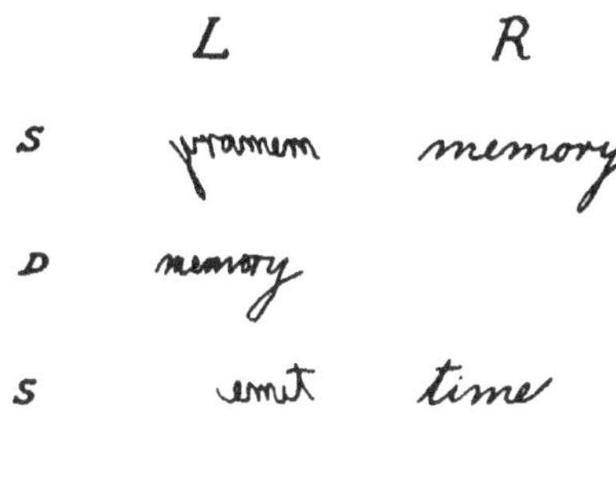

Fig. 180. Symmetrical and Direct Cross-memory.

Symmetrical memory.

A most curious fact about this cross-memory is that the memory for movements is symmetrical and not identical. We learn to write with the right hand; when we attempt to write with the left we succeed fairly well by writing outward (*i. e.*, backward), just as the right hand wrote outward, but we cannot write as well in the regular direction. Here are two specimens (Fig. 180). By looking at the words with a mirror it will be seen that with the left hand those written outward are better than those written inward.

Law of cross-memory.

Some experiments, not extended far enough to enable me to put the law in a quantitative statement, seem to indicate its general forms as follows: The average change produced by cross-memory is composed of two parts, that due to the crossing and that due to the interval of time; the average uncertainty is always much greater than in memory without crossing and increases much more rapidly.

Method of experiment.

The method used in these experiments was the same as that used on page 240. The original line was drawn with one hand, and was repeated with the other, alternately symmetrical and direct. In the particular set of experiments referred to the results were as follows: The remembered line was, on an average, sixteen per cent shorter in the symmetrical movement and twenty-four per cent shorter in the direct movement. The average uncertainty was nine per cent in the symmetrical and nine per cent in the direct.

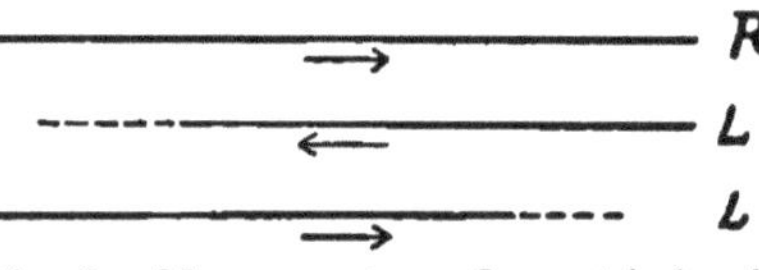

Fig. 181. Measurements on Symmetrical and Direct Cross-memory.

These results can be indicated as in Fig. 181. The top line is the standard, drawn by the right hand in the direction of the arrow. The two other lines are averages of those by the left hand; the portions in dashes indicate the regions within which these lines ended. The irregularity is the same for both, but although both movements differ from the standard, the unsymmetrical one is the less correct of the two.

Memory for tones.

Memory for tones can be measured in a similar way to that employed on pages 139, 140, in determining the least noticeable difference. In fact, all the experiments on the least noticeable difference might be considered as experiments on memory with a very small interval of time between the two impressions compared. There we used an interval of two seconds; by changing this interval to five seconds, ten seconds, etc., we get the record of the size of the least noticeable difference as depending on the lapse of time. The matter is so simple that further explanations hardly seem necessary. A beautiful set of experiments might be performed with the tone-tester, described on page 141.

The method of percentages of correct answers may also be employed. The experiments described on pages 143, 144 are to be repeated with different intervals.

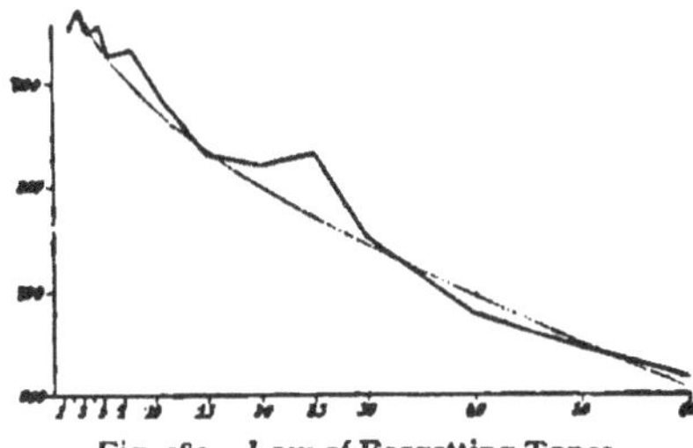

Fig. 182. Law of Forgetting Tones.

Results.

The results of an investigation of this kind are shown in Fig. 182. Here the figures on the horizontal line indicate the number of seconds that elapsed between two tones to be compared, and those on the vertical line indicate the percentages of right answers.

It is seen that the maximum certainty is reached at two seconds. Thereafter it decreases. At an interval of sixty seconds the uncertainty is so great that the answers are nearly half right and half wrong; since mere chance would make them half right, the uncertainty is almost complete.

Individual differences.

This is a characteristic case for many unmusical persons. Individuals differ, of course. There are intelligent persons who cannot recognize a tone repeated twice in close succession. On the other hand, we find Mozart and later piano-players who can carry in mind the slightest differences. Probably the most accurate tone-memory on record is that of Mozart. Two days after playing on a friend's "butter-fiddle" (as he called it on account of its soft tone), the seven-year-old Mozart, while playing on his own violin, remarked that the butter-fiddle was tuned to half of a quarter of a tone lower than his own. And this was found to be the case.

Memory is no real process.

We might make similar experiments on touch, temperature, smell, etc. In fact, memory is no real process; it is merely a way of considering and comparing two impressions at different times. This is what we did with a small interval on many occasions in the earlier chapters of this book. When the interval is so small as to be negligible we speak of simultaneous impressions.

Statistics on memory.

When a man has no brains to invent methods of exact measurement he falls back on statistics; and these very same negative brains assist him in making his statistics worthless. It is the fashion to collect statistics on memory, but only two really scientific investigations of this sort have ever been made.

Great sources of error.

Numerous sets of calculations of the number of letters or words forgotten out of the total number seen, heard, spoken, etc., have been undertaken. Letters and words are very complicated affairs, and the results will vary completely by a slight change in the word, in the arrangement, in the time, in the loudness or illumination, in the intonation or the size, etc., etc. The sources of error are so great that a scientist, *i. e.*, a careful worker, must spend years of labor in getting them under control.

Law for memory of syllables. The first carefully executed experiments in this line show that when a set of meaningless syllables has once been learned, the time required for learning them on a second occasion increases as the interval between the two occasions, according to a definite law.

This law runs in the way shown in a specimen table of results:

Interval . .	0.3	1	8.8	24	48	144	744 hours
Per cent of work required for relearning	42	56	67	66	72	75	79

At first there is a rapid loss, more than half during the first hour ; then the loss is steadily less rapid and finally becomes almost steady. Between the second day and the thirty-first day there is almost no change.

Further experiments with letters under various conditions of rate, repetition, lapse of time, rhythm, etc., have been in progress for many years, but the final results have not been reached.

Concrete cases. Statistical experiments require an immense amount of labor, and seldom lead to satisfactory results when employed to determine fundamental laws of mental life. The case is quite different when the question to be answered applies to a single concrete case. The question of how much the boys of a class have remembered from the last lesson, twenty-four hours ago, can be answered with more or less accuracy by an examination. The determination of a general law of memory in such a manner that knowledge of certain circumstances enables the prediction of how much will be remembered at any future time is another matter altogether.

Prodigies of memory.

The education of the memory powers has ever been a subject of interest to practical people. More or less fabulous accounts of the prodigies of memory may be found in various psychological story-books. Even when the records of the results obtained are to be credited, the accounts of how the freaks educated their memories are mostly to be regarded as unconscious fiction. For practical purposes statements on the development of memory should be founded on observation of and experiment on ordinary people.

Fundamental laws for educating memory.

The fundamental laws for the cultivation of memory are: intensifying the image by attention, and keeping it ready by conscious repetition.

In the first place, intensify the impression. See, hear, do what you wish to remember. You cannot expect to remember a picture when you have not really seen it. It is said that the Nürnbergers never hang a man till they have caught him, and yet many a teacher expects his pupils to remember a lesson without really learning it.

Methods of intensifying the impression.

How shall we intensify the impression? Any method that increases the amount of attention will help to intensify the impression; these methods have been considered in Chapter VII. But it is not sufficient merely to pay attention; something further must be done if the impression is to be retained. No experimental work in the laboratory has been done on this problem, but some of the most acute experimenting has been carried on by advertisers on account of the business interests involved. The very principles they have discovered are just the ones we should make use of on ourselves and in teaching others. I believe that these principles have never been formulated and that advertisers follow them unconsciously—we can walk successfully although we may know nothing about the action of the muscles of the leg.

First principle. A powerful principle employed for memorizing a fact is that of the ridiculous. You cannot forget the absurd pictures by means of which publishers and players advertise their new wares; or Paderewski's hair, whose echoes lasted longer than those of his playing; or the tramp army, whereby "General" Coxey hopes to live in history.

A subordinate principle belonging to the ridiculous is that of the pun. A good pun is an æsthetically ridiculous contradiction; a bad one is intensely irritating but is ridiculous ridiculousness. If you wish your class to remember the story of Waterloo, make a pun about it, and a bad one, too. (You all know the horrid one to which I refer.)

THROUGH
YOU CAN GET "A HOLD" ON THE PEOPLE

Fig. 183. Use of the Pun for Memory Purposes.

Second principle. A second principle of memorizing is that of rhyme. We all know how much easier it is to learn rhymed poetry than blank verse or prose. Rhymed couplets or verses can frequently be employed to memorize difficult facts. The farmer's calendars in olden times were based on the memorial days of the saints. To remember when the sowing, reaping, etc., should be done, an appropriate couplet was rhymed with the day. The same method is employed in some aids to learning history. Those who have studied formal logic will remember the medieval memory-verse beginning, "Barbara, Celarent," etc. Students of medicine are required to know the names and arrangement of the bones in the hand. Being a very difficult matter, the German students have been ingenious enough to make a translation of the Latin names into an absurd stanza. Over six years ago I happened to hear this a couple of times; to-day the lines are still in memory:

Combination of the two principles.

"Vieleckig gross, vieleckig klein,
Der Kopf muss bei dem Haken sein.
Dann schiffen wir beim Mondenschein
Dreieckig über's Erbsenbein."

The chief words when translated into Latin give the names required.

This principle of rhyme when combined with the ridiculous can be carried so far that couplets and stanzas *cannot* be forgotten. Those who have read Mark Twain's story about "Punch, Brothers," etc., will remember a case. In order to spare a very disagreeable experience to those who have not been haunted by this stanza, I will not repeat it.

Third principle.

The principle of alliteration, *i. e.*, of words beginning with the same sound, was largely used in olden poetry. Memory was doubtless greatly assisted thereby. It is in use to a certain extent to-day in book-titles, catch-words, advertisements, etc. Sometimes it is used unintentionally. The nation will never forget the famous phrase of the presidential campaign of 1884, "Rum, Romanism, and Rebellion."

Fourth principle.

Another very efficient principle is that of puzzle. Dissected maps, the game of authors, the solution of mathematical conundrums, are cases.

Second law.

To retain things in memory they should generally be repeated a number of times. With a very intense first impression the repetition may be unnecessary ; with weak impressions it may be frequently required. The relation of intensity to repetition has, however, never been experimentally determined.

Emphasis on "conscious."

The fundamental fact to be observed is that the repetition must be conscious. Nearly everybody supposes that a series of facts, a group of names, a collection of dates, can be learned by simple mechanical repetition. It is not too strong to say that "learning by rote" is an absolute impossibility. We remember the connection between two words when we pay attention to the fact of such connection. For example, suppose we wish to remember that Aristotle was a tutor to Alexander. The fact strikes us at once and will have some power of persistence in our memories. Any amount of mechanical repetition of "Aristotle-tutor-Alexander" will not assist. But let each repetition be a conscious, attentive connection of the three facts ; there is a distinct gain. The difficulty lies in making the repetition conscious and not mechanical.

Two methods.

The methods of doing this may be described as voluntary and involuntary. In the voluntary method the individual calls up each time by an effort of will a characteristic picture of Aristotle teaching Alexander. The involuntary method consists in finding some word naturally connected with Aristotle which by another natural connection brings up another word and so on till "teach" is reached, after which the same process stretches from "teach" to "Alexander." Teachers of memory-culture, like Loisette, have made a special application by the method of searching for a series of connecting associations between the two words or facts to be remembered.

The objection made to such associative systems is that they are too cumbersome when anything is to be recalled. While practicing with one of these systems I noticed the tendency of the middle links to fall out; no matter how many intermediate words were inserted between "Aristotle" and "teach," after a while the two were involuntarily associated, with no thought of the middle links. This process, which is in harmony with facts previously discovered concerning the association of ideas, might be called the obliteration of intermediate associations.

Law of obliteration of intermediate links.

Like all our mental life, memory depends upon age. In a series of experiments on school children a tone was sounded for two seconds, then it was started again and the child was required to stop it when it had lasted as long as before. In all cases the second sound was made too short; the younger children often made the sound by memory only one fourth of its true length. As they grew older, the memory became more accurate.

Dependence of memory on age.

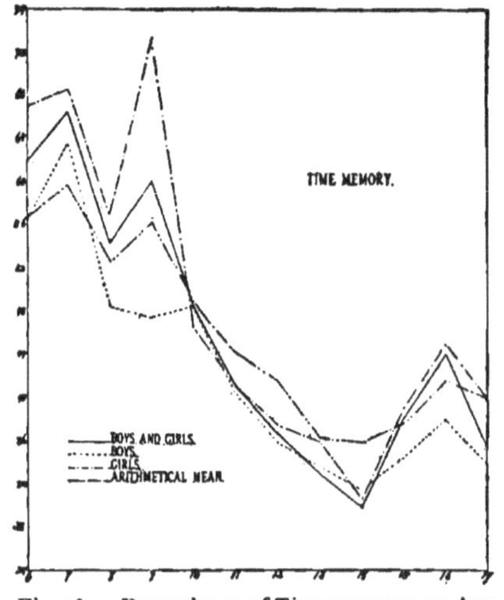

Fig. 184. Dependence of Time-memory on Age.

Concerning the ages above seventeen no experiments have been made. We know, however, that old people

Memory and old age.

gradually lose their memories. Indeed, we might say that memory is the ostensible friend who insists upon presenting us with a house bountifully furnished with the skeletons of past sins, but who in old age turns us out into the cold night of forgetfulness when we would gladly remember even the sins. Memory grows to its prime and then never gets any further till it suddenly becomes withered and past.

CHAPTER XIX.

RHYTHMIC ACTION.

WHAT is rhythmic action? Such a hard Greek word as "rhythm" (alas! there is no English word) must mean something very dreadful. Do you remember M. Jourdain in Molière's "Le Bourgeois Gentilhomme," who was astounded and delighted to learn that he had been speaking "prose" all his life? Well, you have been executing rhythmic actions ever since you began to walk. **Definition.**

By rhythmic action we understand an act repeated at intervals which the doer believes to be regular. Walking is in simple rhythm. The beating of a drum is intended to be in a more or less complicated rhythm. **Examples.**

Let us take a lesson in walking. In order that there may be no dispute on the subject and that we may have a permanent record, we shall try to arrange matters so that every step is recorded. While studying action we learned the principle of graphic recording by air transmission; all we have to do now is to modify the method so that it records the movement of walking. **Walking.**

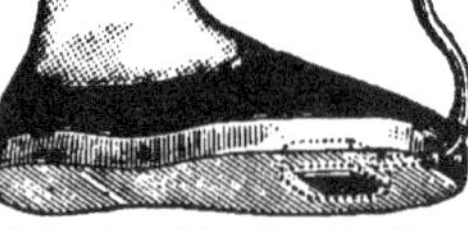

Fig. 185. The Pneumatic Shoe.

The person experimented on puts on a pair of shoes with hollow rubber soles (Fig. 185) which act as receiving capsules. Each sole communicates by a long tube with a small capsule that writes on a small smoked drum (Fig. **Graphic method applied.**

186) carried in the hand. When the foot is on the ground, the air is pressed through the tube to the recording capsule ; this causes it to make a mark on the drum.

Results.

The character of the results is indicated in Fig. 187. The length of time during which the foot rests on the ground is indicated by the length of the mark on the drum. In walking, one foot leaves the ground just as the other touches it ; in going upstairs, both feet touch for a while at the same time ; in running, both feet are off the ground for short intervals.

Fig. 186. Walking with Pneumatic Shoes and Recording Drum.

But all this was already known in sporting circles ? Still, you must not object to putting a competitor or even an umpire on record. In a walking-match a man is ruled out by the umpires if his method of progress changes from 1 to 3 (Fig. 187). What a lot of quarreling would be saved if every man could carry on his back a minute instrument telling his walk in black and white !

Fig. 187. Graphic Records: 1, Walking; 2, Going upstairs; 3, Running; 4, Faster Running.

Psychology of walking.

The interest of the physiologist ends where that of the psychologist begins. The physiologist knows that we walk with our feet ; the psychologist knows that we walk with our minds also. We *will* to walk faster or slower, this way or that ; how does the execution compare with the intention ?

The method just described was developed for physiological purposes and has not been used for a study of the psychology of walking, although that could be done with very little trouble.

Electric shoe.

For a study of the influence of the mind on walking I have devised a little reaction-key for the foot, to be used with the spark method. This key is shown in Fig. 188. It is attached to the heel of the shoe; flexible conducting cords lead from it to the spark-coil. The spark-coil is arranged to record on the drum by making a dot on the smoked paper. The rest of the arrangement depends on the particular question to be studied.

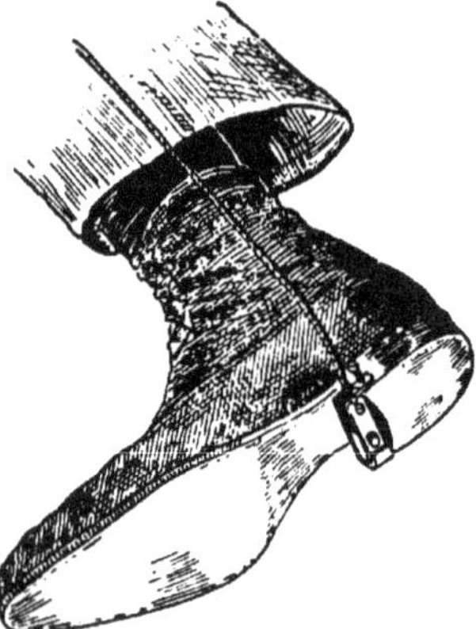

Fig. 188. The Electric Shoe.

Marking time.

MARK TIME! Left, right, left, right, etc. The drum beats rub-a-dub-dub and Sergeant Merritt at the end of the line brings his foot down exactly in time with the strokes of the drum. Yes, *exactly* in time. Sergeant Merritt is not an ordinary sergeant; his is the crack company of the Seventh Regiment. The whole world knows that everything is *exactly* right in that regiment, and nothing short of a stroke of lightning would convince the sergeant that he is behind time. Let us try our spark method, which is merely lightning on a small scale. But before we begin an experiment we must distrust everything and everybody—even the drummer. The drummer himself may have something the matter with him—we will attend to that later—but at any rate we must use some arrangement for drumming which we have proved to be exact.

Preparations for experiment.

The drumming we shall use will be a series of clicks at exactly equal intervals. To produce the click we use the graphic chronometer. This is essentially a stop-watch which makes the fine pointer beat either in seconds or in fifths of a second. This pointer writes on the smoked drum. At the same time it breaks an electric current and makes a click by means of a telegraph sounder.

A foot-key is fastened on one of the sergeant's heels and the wires are led to the spark-coil, just as in the case of the piano-player (Fig. 6). The sergeant's case is not that of simple reaction to sound ; he knows, from memory of time, just when the clicks are coming.

The record on the drum will be like that shown in

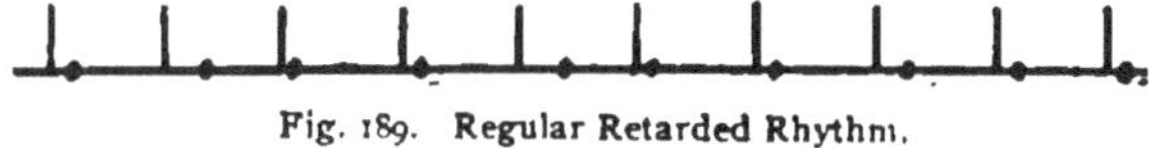

Fig. 189. Regular Retarded Rhythm.

Fig. 189. It shows a line drawn by the chronometer point, on which, at regular intervals representing seconds, there are side lines denoting the moments of the clicks. The dots are made by the sparks at the moment the heel touches the floor.

Regular retarded action.

The sergeant is, alas! always just about one sixth of a second behind time. He is very regular about it, too ; for he is a rather stolid, unexcitable fellow on whom you can depend for "getting there," although he may not be so lively as another.

Irregular retarded action.

When the sergeant saw his record, it worried him into making an effort at being on time. His second record was like Fig. 190.

Gained he had not ; on an average he was as much

behind as at first. But his nervousness had added a worse fault, that of irregularity.

Regular accelerated action.

Next to the sergeant comes Corporal Alan Adair,

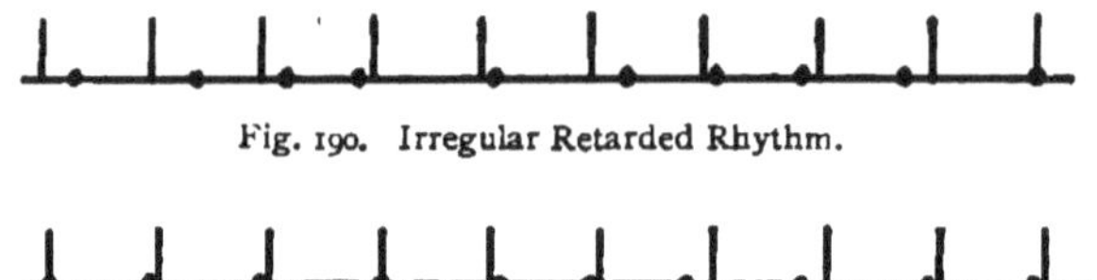

Fig. 190. Irregular Retarded Rhythm.

Fig. 191. Regular Accelerated Rhythm.

eager and enthusiastic. He always speaks before he thinks ; his record shows that in his ardor he gets quite ahead of the drum (Fig. 191).

Irregular action.

We have also in our company a Frenchman, Antoine Boulanger. His record (Fig. 192) proves to give a good average, but it is very irregular. Antoine, we all know, is a first-rate fellow, although he is inclined to be very nervous and excitable.

All the persons tested show records that can be classi-

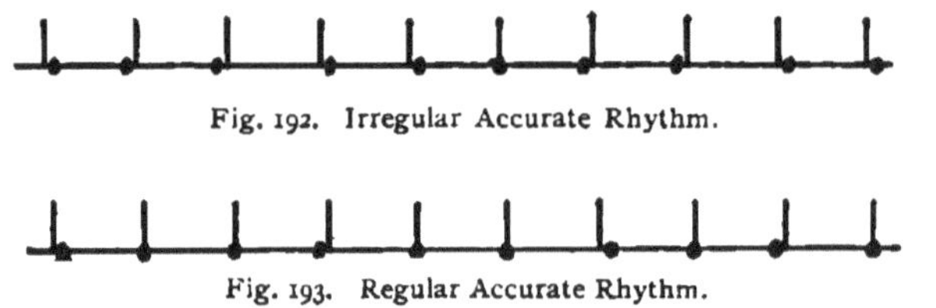

Fig. 192. Irregular Accurate Rhythm.

Fig. 193. Regular Accurate Rhythm.

Regular accurate action.

fied on two principles, accuracy and regularity. Accuracy is the nearness of the general average to the series of clicks. Regularity is the person's agreement with himself. A man may be accurate but irregular, like Antoine, or inaccurate and irregular, like the sergeant when nervous, or inaccurate but regular, like Alan. Finally, when the foot comes down always within a small range before and after the click, so that it, on an average, hits the click (Fig. 193) the record is both accurate and regular. This is the ideal of rhythmic action.

Teachers can readily pick out the very bad cases of inaccuracy or irregularity among a class of marching boys. Drill sergeants can tell tales of their raw recruits.

Computing the results.

The distance between each two of the checks in the preceding figures means an interval of one second. With a fine measure, or even by the eye alone, we can divide the interval into ten parts, each of which will mean one tenth of a second. Now, note down how many tenths of a second the dot is distant from the check; if it is ahead of the check, put + in front of it; if behind, —. The record in Fig. 190, for example, will be

$$-3, -4, -2, +1, -1, -3, -1, +1, +2, 0.$$

Index of inaccuracy.

Take the average, that is, add them all up and divide by ten. This gives —1.0 tenths of a second as the average amount by which the foot was behind time. In physics this is called the constant error; in psychology—especially in educational psychology—I propose to call it the "index of inaccuracy."

Index of irregularity.

Now let us find the "index of irregularity," or, as physicists call it, the variable error. Find the difference between the number in the index of inaccuracy, in this case 1, and each of the numbers, 3, 4, 2, 1, etc., of the original records. You will get a second set of ten figures, 2, 3, 1, 0, 0, 2, 0, 0, 1, 1. As you will notice, no attention has been paid to + and —. Average these last results; answer, $\frac{10}{10}$, or 1.0, of a tenth of a second, which is the index of irregularity. By chance the two indexes have the same figures.

A very irregular person might have the same index of

accuracy as a very regular one ; they might both be one tenth behind time ; but their indexes of irregularity would be different. On the other hand, two regular persons will have small indexes of irregularity, whereas their constant errors would be quite different.

Experiments on the drummer.

Now, to attend to the drummer. Suppose we put into his hand an electric drum-stick. Every time that the stick strikes the drum a spark is made. Since the drummer has no watch to guide him but judges his time as he pleases, we do not use any sounder but let him beat alone. A record can be made just as before with the chronometer, and the regularity can be measured in tenths of a second.

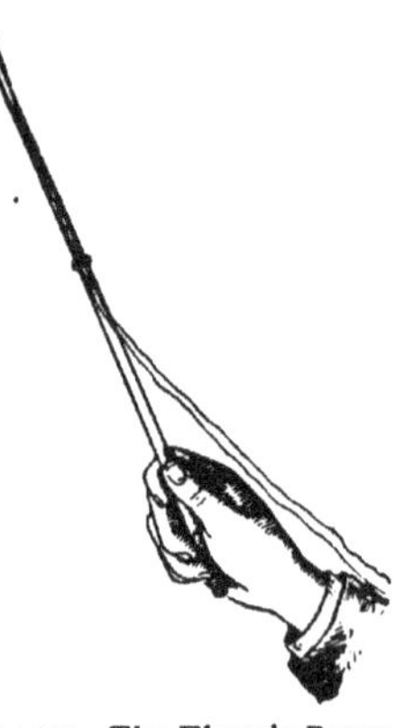

Fig. 194. The Electric Baton.

The orchestra leader.

The index of irregularity is of immense importance to the orchestra leader ; there is no index of inaccuracy, because he sets his own time. It does not make much difference just how *fast* he beats, provided he beats *regularly*. To measure the regularity in a case of this kind an electric contact on the end of a baton can be arranged by which a spark record is made in the usual way (Figs. 194, 195).

Example.

The time between each record can be measured in hundredths or thousandths of a second, as desired. Suppose we have a record of eleven beats measured to hundredths of a second with the following results : 41, 42, 37, 41, 39, 40, 40, 40, 41, 38, 41. The average time of a beat is just 40. How *regular* is the beating? This is determined by finding the difference between each sepa-

Fig. 195. Taking an Orchestra Leader's Record with the Electric Baton.

rate beat and the average, and taking the average of these differences. Thus:

41	1
42	2
37	3
41	1
39	1
40	0
40	0
40	0
41	1
38	2
41	1
11 440	11 12
40	1.1

The index of irregularity is 1.12.

Now let us take *another* orchestra leader whose record gives 40, 41, 42, 40, 39, 37, 35, 40, 41, 41, 38; which is the better man? The average is 40 as before, but the index of irregularity is 1.8 as compared with 1.1.

An irregular leader.

Suppose we have a *third* leader from whom we get the ten records: 40, 39, 40, 40, 39, 38, 39, 39, 39, 39. The average is 39.2, and the index of irregularity is less than 0.5.

A regular leader.

It is evident that the second leader beats so irregularly that an orchestra cannot possibly keep time, that the first leader is somewhat better, and that the third is far superior to the others. The actual *average* time of a beat makes no difference within such small limits, as music played at the rate of one beat in 0.40 seconds is not sensibly different from that played at one beat in 0.39 seconds. An essential qualification, however, for the success of an orchestra leader is his *regularity* in estimating intervals of time.

Piano-playing.

Another example similar to the one just mentioned is that of a piano-player, who must learn to strike the notes at regular intervals. The quarter-notes should all be about the same length; equal measures should be completed in equal times. For most beginners the irregularity in the time given to successive measures varies to such an extent that it is painful to hear them attempt a tune. By practice with the metronome successful players are able to reduce their irregularity till it does not disturb the playing. It is not known just how far this may be carried, as no one has ever taken the trouble to make measurements. It might be suggested, however, that, even when the irregularity is so small that no one notices it, yet it may be great enough to injure the effect. A successful musician of any kind should know not only that his *instrument* is in *tune* but also that he *himself* is in *time.*

Dumb-bell exercises.

The rhythmical exercises with dumb bells are the expression of an instinctive desire of the gymnast to cultivate his accuracy and regularity of action. To make the measurements a flexible wire is inserted into the handle of each of a pair of iron dumb bells and is connected with the spark-coil so that when the metal ends are struck together a spark is made. Front and back movements

Fig. 196. Taking a Record with Electric Dumb Bells.

(or the rataplan) are well adapted to measurements.

Final analysis of rhythmic action.

What is rhythmic action? The process in the mind of the one who is acting is in the first place an estimate of equal intervals of time; after a few strokes at equal intervals the person knows just when to expect the next one. In other words, it is a case of time-memory corrected by an actual stroke each time. Knowing when to expect the next stroke, an act of will is executed so that the final action occurs in some definite relation to the stroke, generally at the same moment or just after it. This process might be called a reaction to an expectation. In extreme cases the act of will may be so late that the action seems actually a reaction to each stroke. This would be the case with the sergeant. In some cases of congregational singing the leader keeps about a quarter of a second ahead of the congregation, implying that they are incapable of singing the tune and must rely on reaction to each note as heard. Such réactions are, however, so complex that this method could hardly be of use unless the leader is very far ahead.

CHAPTER XX.

SUGGESTION AND EXPECTATION.

A suggestion from the time of day.

IN HIS memoirs Robert-Houdin begins with a description of the effects of suggestion from the time of day.

"Eight o'clock has just struck: my wife and children are by my side. I have spent one of those pleasant days which tranquillity, work, and study can alone secure—with no regret for the past, with no fear for the future, I am—I am not afraid to say it—as happy as man can be.

"And yet, at each vibration of this mysterious hour, my pulse starts, my temples throb, and I can scarce breathe, so much do I feel the want of air and motion. I can reply to no questions, so thoroughly am I lost in a strange and delirious reverie.

Originated by associations.

"Shall I confess to you, reader? And why not? for this electrical effect is not of a nature to be easily understood by you. The reason for my emotion being extreme at this moment is that, during my professional career, eight o'clock was the moment when I must appear before the public. Then, with my eye eagerly fixed on the hole in the curtain, I surveyed with intense pleasure the crowd that flocked in to see me. Then, as now, my heart beat, for I was proud and happy of such success.

"Do you now understand, reader, all the reminiscences this hour evokes in me, and the solemn feeling that continually occurs to me when the clock strikes?"

Vague use of the term.

The effect of suggestion—what has not been included under this term! According to some of the hypnotismus "psychologists," all mental life from the simplest impressions of the senses up to the highest creations of art and social life—all is nothing but suggestion.

Errors of the "hypnotismus" psychology.

Volumes upon volumes have been written on hypnotism and suggestion; indeed, the list of works itself fills a volume with over 2,000 titles. But at the end of it all, what have we besides careless observation, vague guesswork, and endless speculation? It is all on the level of the old psychology, not an experiment in it.

Perhaps the most curious point in the case is that among the hypnotism dilettants, the mesmeric mysticists, and the long-winded double-consciousness researchers there is actually a society for experimental (!) psychology. Most of these people have duped themselves into the belief that they are contributing to science; this ceases to be self-delusion and becomes deliberate swindle when they mislead the public by calling their inanities by the term "experiment."

Possibility of experimenting on suggestion.

But why should it not be possible to experiment on suggestion? Why should it not be possible to actually measure a suggestion and its effects? It *is* possible. As in all new undertakings, the way was hard to find; our attempts have cost endless thought and labor, and we have a choice collection of failures as mementos. But when we have found the way, it seems strange that we and everybody else were so blind as not to see it long ago.

What has been accomplished.

Yet, not too much must be expected. The method by which we have measured the suggestive effects of size on weight will be explained and the results will be given. This will serve to give a general idea of one of our methods. What I cannot do here is, to give an account of the extended researches on suggestion and hallucination that have been carried on in my laboratory during the last two years. We have found the way to measure in so many candle-power a suggestive effect of sight, we can produce hallucinations of tones that are equal in intensity to real tones whose physical energy can be measured, we can cause a person by walking a certain number of feet to see a spot where there is none. These experiments have been ably and patiently executed under my directions by one of my pupils, but it is the rule for such investigations to appear first in the "Studies from the Yale Psychological Laboratory."

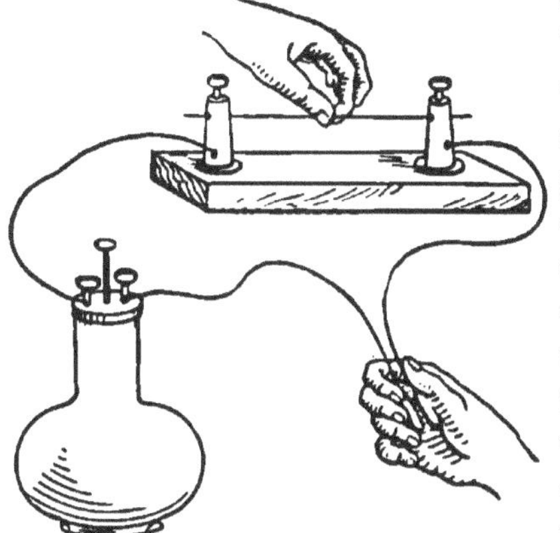
Fig. 197. Producing an Hallucination of Warmth.

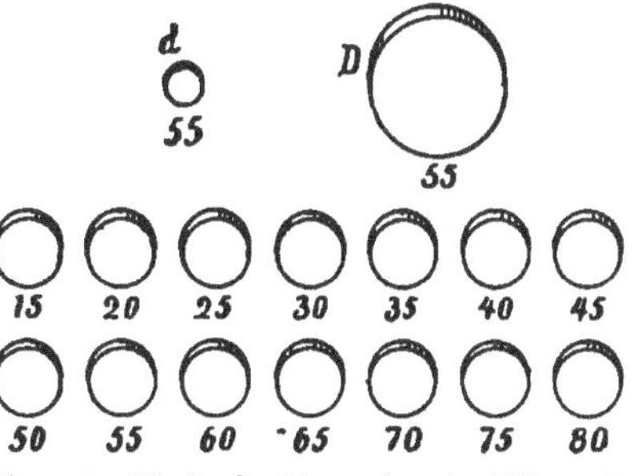

Fig. 198. Blocks for Measuring the Effect of a Suggestion of Size.

The suggestion blocks. Here is a series of round blocks painted black; in appearance they are all just alike, but in weight they are different. This block *D* is a very big block ; pick out that one of the series which appears of the same weight as the big one, when lifted between thumb and finger. You know nothing about the blocks except that, to the best of your belief, the big one is of the same weight as the medium-sized one. Put them on the scales ; down goes the big one, you judged it to be much lighter than it was. Try it over again as often as you please ; always the same result. By means of the scales find the medium one that weighs exactly the same as the large one. Then compare them by lifting ; nothing but the incontestable evidence of the scales will make you believe they are the same. After being familiar with the experiment for over a year I still find the effect almost as strong as at first.

Making the experiment.

But how much? It is not sufficient to show that there is a suggestive effect, you must measure it. The difference in weight between the two blocks supposed to be equal gives the effect of suggestion in just so many ounces or grams.

How much?

In a set of experiments carried out on school children the medium-sized blocks were graded in weight from 15 grams to 80 grams. A large block *D* and a small block *d*, each of 55 grams, were successively compared with the set of graded blocks. The difference between the weight picked out for the larger one, *e. g.*, 20 grams, and that for the smaller one, *e. g.*, 70 grams, would give the effect of the difference in size between the two blocks. The difference in weight in this example would be 50 grams, which would be the result of the difference of six centimeters in the diameter of the blocks.

Experiments on school children.

The effect of the suggestion depends upon the age. The results for the New Haven school children are indi-

Dependence on age.

cated in Fig. 199. The figures at the bottom indicate the ages; those at the left the number of grams in the effect of suggestion.

About 100 children of each age from 6 to 17 were taken. The average effect of the suggestion was as follows: 6 years, 42 grams; 7 years, 45 grams; 8 years, 48 grams; 9 years, 50 grams; 10 years, 44 grams; 11 and 12 years, 40 grams; 13 years, 38 grams; 14 to 16 years, 35 grams; 17 years, 27 grams. For all ages the

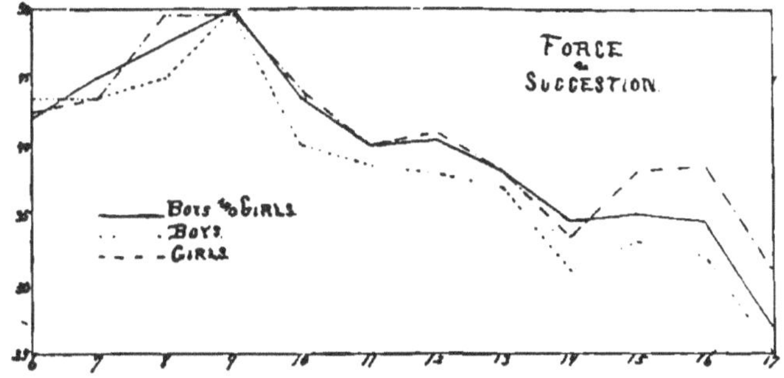

Fig. 199. Dependence of the Effect of Suggestion on Age and Sex.

average was above twenty-five grams. The suggestibility slowly increases from six years to nine years; after nine years it steadily decreases as the children grow older. The results, when separately calculated for boys

Dependence on sex.

and girls, show that at all ages the girls were more susceptible to suggestion than the boys, with the exception of the age nine, where both were extremely susceptible.

Extreme cases.

These are the average results for large numbers of children. Many young people, however, were so susceptible that the set of middle-sized blocks did not range far enough to suit them. At the age of seven years 37 per cent of the children declared that the large block was lighter than the lightest block, and that the small block was heavier than the heaviest. The actual difference between them was 65 grams ; thus the effect of suggestion was more than the weight of the suggesting blocks *D* and *d*.

Factors in the suggestion from size.

The factors that produce such a deception of judgment seem to consist in a suggestion—or, rather, a disappointed suggestion—of weight. Big things are, of course, heavier than little things of exactly the same kind. When we find two things of the same appearance but differing in size, the big thing *must* be heavier. This reasoning is all done without our suspecting it, and we unconsciously allow our judgment of weight to be influenced by the size as seen. When the eyes are closed and the blocks are lifted by strings, of course there is no illusion.

A pound of lead and a pound of feathers.

Which is the heavier, a pound of lead or a pound of feathers ? A pound of lead, says the unsuspecting person, and then you guy him for his stupidity. But this poor fellow, who has been laughed at for centuries, is right. A pound isn't a pound all the world over ; it all depends on how the pound looks. A pound of lead is heavier than a pound of feathers. Try it with a pillow and a piece of lead pipe. No matter if the scales do say that they weigh just the same, the pound of lead is much the heavier as long as you look at it.

Suggestion from the span of the fingers.

In the preceding case we have had a suggestion from sight alone. Similar effects are produced by differences in the span of the fingers. Suppose we have all our blocks of exactly the same diameter. We have one set just alike in size but differing in weight, and other blocks of just the same diameter and weight but differing in length, one being very long and the other very short. The experiments are made in the same way as before except that the eyes are closed. The suggestion arises from the difference in span of the fingers for a long block and a short one. By looking at the blocks with the eyes open, a sight-suggestion is added to the muscular suggestion.

Suggestion by movements; In the preceding cases it has been noticed how a suggestion causes a change in judgment; there is another field in which suggestion is very active, namely, the suggestion of movement. While a person is exerting his whole power on a dynamometer (page 83), let him observe contracting movements of your hand. He soon feels irresistible twitchings in his own hand and actually exerts still more force.

The suggestion of movement may even take effect against the will of the person concerned. A child in school with the Vitus dance will sometimes be involuntarily imitated by the others. A contagion of this kind that occurs in every-day life is the effect of gaping.

by gesture; The orator and the actor make use of expressions and gestures intended to arouse similar impulses in their hearers and consequently to make their ideas more effective.

by expression of the face. On the other hand, if you wish to get at the thoughts of a person with whom you are speaking, you should look steadily at his face. His expression cannot help changing, and these changes produce similar changes in your own face, thereby awakening various emotions of doubt, confidence, anxiety, etc. This was an art of old-time diplomacy. The readiness of women to read characters in this way may be due to their greater susceptibility to suggestion.

Irresistible suggestions. Every idea of a movement brings an impulse to movement. This is especially prominent in those rare individuals who cannot keep a secret. The very reading and thinking about crimes and scandalous actions produce a tendency to commit them. In some persons this influence is quite irresistible. As soon as one bomb-thrower attacks a rich banker, everybody knows that within a week half a dozen others will do the same. No sooner does one person commit suicide in such a way that it is strikingly described in the newspapers, than a dozen others go and do likewise.

The runner's trick of suggestion.

A runner, prepared to start, can often cause the starter to fire his pistol unintentionally by starting to run. This runner is ahead of the starter by the amount of the starter's reaction-time, while the other runners are behind the starter by the amounts of their own reaction-times. As the reaction-time may readily amount to one third of a second, the runner who relied on the suggestion may gain by a large fraction of a second.

Significance of suggestibility.

Thus we have gathered a few facts on suggestibility by experiment. The full significance of suggestibility is apparent when we remember that teaching, preaching, acting, public speaking, and pleading are forms of suggesting. The freaks of hypnotism are performed by suggestion. The faith-cures and the miraculous effects of the Grotto of Lourdes are benevolent suggestions. The ceremonials of our churches are suggestions bringing us into a religious frame of mind. The manipulations of the spiritualists and the monotonous blackness of a funeral are all forms of suggestion. How shall we develop the children so as to produce in them minds well balanced in respect to suggestion? Is this not as important a task as learning to do percentage or to parse a sentence? Here is a field where the educator and the psychologist must seek for facts.

Suggestive expectation.

In expecting an event we have some thought in mind; this thought often acts as a suggestion.

Effect on reaction-time.

The time of reaction depends on its expectedness; unexpected events require in general more time and produce very irregular results. It is customary to give a warning click about two seconds before an experiment. Experiments on one person give a reaction time of 305 without warning and 188 with warning.

Different direction of attention.

It also makes a difference if the attention is directed to the stimulus expected or to the movement to be executed. In general the latter method is quicker, but with some persons the reverse is the case. Experiments made on one subject give as reaction-time to sound the result 216 when the attention was directed toward the expected sound, and 127 when it was directed toward the finger to be moved.

Effect in astronomical records.

The expectation that a star will pass one of the hair-lines in a telescope produces differences in regard to the time of its passage as actually recorded. This phenomenon, which led to the discovery of mental times, is more complicated than the simple cases of reaction-time and thinking-time that we have considered in Chapters III. and IV.

Passage of a star.

Let me illustrate how this happens by a simple case. Suppose that we have to determine the time of the passage of a star at some distance from the pole across the meridian. We may employ an old astronomical method which is still sometimes used for time-determinations, and which is called the "eye and ear method." A little before the time of the expected passage, the astronomer sets his telescope, in the eye-piece of which there have been fixed a number of clearly visible vertical threads, in such a way that the

Method of recording.

middle thread exactly coincides with the meridian of the part of the sky under observation. Before looking through the instrument, he notes the time by the astronomical clock at his side, and then goes on counting the pendulum-beats while he follows the movement of the star.

Estimation of fractions of a second.

Now the time-determination would be very simple if a pendulum-beat came at the precise moment at which the star crosses the middle thread. But that, of course, happens only occasionally and by chance ; as a rule, the passage occurs in the interval between two beats. To ascertain the exact time of the passage, therefore, it is necessary to determine how much time has elapsed between the last beat before the passage and the passage itself, and to add this time—some fraction of a second—to the time of the last beat. The observer notes, therefore, the position of the star at the beat directly before its passage across the middle thread, and also its position at the beat which comes immediately after the passage, and then divides the time according to the length of space traversed.

Influence of the observer's attention.

If *f* (Fig. 200) is the middle thread of the telescope, *a* the position of the star at the first beat, and *b* at the second, and if *af* is, *e. g.*, twice as long as *fb*, there

Fig. 200. Actual Positions of the Star at the Pendulum-beats.

must be added ⅔ of a second to the last counted second.

It has already been told (page 40) how the astronomers disagree in their records although the star would have the same position for all. A constant and regular difference, such as this actually is, can be explained on the assumption that the objective times of the actual events and the times of their notice by the observer are not identical, and that these times show a further difference from one another according to the individual observer. Now, attention will obviously exercise a decisive influence upon the direction and magnitude of such individual variations. Suppose that one observer is attending more closely to the visual impression of the star. A relatively longer time will elapse before he notices the sound of the pendulum-beat. If, therefore, the real position of the star is *a* at the first beat and *b* at the second (Fig. 201) the sound will possibly not be noticed till *c* and *d*, so that these appear to be the two positions of the star. If *ac* and *bd* are each of them ⅓ of a second, the passage of the star is plainly put ⅓ of a second later than it really should be.

Visual attention.

Auditory attention.

On the other hand, if the attention is concentrated principally on the pendulum-beats, it will be fully ready and

Fig. 201. Supposed Positions with Visual Attention.

properly adjusted for these, coming as they do in regular succession, before they actually enter consciousness.

Hence it may happen that the beat of the pendulum is associated with some point of time earlier than the exact moment of the star's passage across the meridian.

Fig. 202. Supposed Positions with Auditory Attention.

In this case you hear too early, so to speak, just as in the other case you heard too late. The positions *c* and *d* (Fig. 202) are now inversely related to *a* and *b*. If *ca* and *db* are, say, $\frac{2}{9}$ of a second, the passage is put $\frac{2}{9}$ of a second earlier than it really occurs. If we imagine that one of two astronomers observes on the scheme of Fig. 201, the other on that of Fig. 202—in other words, that the attention of the one is predominantly visual, that of the other predominantly auditory—there will be a constant personal difference between them of $\frac{1}{3} + \frac{2}{9} = \frac{5}{9}$ of a second. You can also see that smaller differences will appear where the manner of observing is the same in both cases but with differences in the degree of the strain of the attention; while larger differences must point to differences like those just described, in the direction of the attention.

Difference in attention.

CHAPTER XXI.

MATERIALISM AND SPIRITUALISM IN PSYCHOLOGY.

System of psychology.

IN THE good old days, now happily gone forever, when psychology belonged to philosophy, we were accustomed to hear of materialistic psychology, spiritualistic psychology, the psychology of Hamilton, the psychology of Hegel, English psychology, German psychology, etc., etc.

Absurdity of anybody's "system."

Nowadays it is just as absurd to speak of anybody's system of psychology as to speak of anybody's system of chemistry. There is one science of chemistry to which all scientific chemists are contributors; there is one science of psychology which all scientific psychologists make their humble efforts to develop. How this has come about I am going to tell by translating a few pages from Wundt's "Vorlesungen über Menschen und Thierseele."*

Early psychology.

"The earliest psychology is materialism. The soul is air or fire or an ether; it remains, however, material, notwithstanding the efforts to lighten and thereby to spiritualize the matter. Among the Greeks it was Plato who first freed the soul from the body, whereby he made it the ruling principle of the latter. He thus opened the path for the one-sided dualism which regarded sensory existence as the contamination and degradation of a purely mental being. Aristotle, who united a wonder-

Plato.

Aristotle.

* The whole work has been translated and published under the title, "Lectures in Human and Animal Psychology."

ful sharpness of observation to his gift of speculation, sought to soften this contrast by infusing the soul into matter as the vivifying and constructive principle. In the animals, in the expression of the human form in repose and motion, even in nourishment and growth, he saw direct effects of mental forces, and he drew the general conclusion that the soul brings forth all organic form just as the artist forms the block of marble. Life and soul were for him the same ; even the plant had a soul. Yet, Aristotle, like no one before him, had studied into the depths of his own consciousness. In his work on the soul, the first book treating psychology as an independent science, we find the fundamental processes carefully distinguished and—as far as possible in his time—explained as to their relations.

Aristotelian psychology.

"The Aristotelian psychology, and especially its fundamental principle that the soul is the principle of life, governed the whole of the Middle Ages. At the beginning of modern times here, as in other subjects, a return to the Platonic views began to weaken its power. Soon a new influence was associated : the revival of the modern natural sciences and the mechanical views of the world which they spread abroad. The result of the conflict was the birth of two fundamental views in psychology which down to the present day have fought each other in the field of science : spiritualism and materialism. Strange to say, the very same man was of primary importance for the development of both. Descartes, no less great as mathematician than as philosopher, defined, in opposition to the Aristotelian psychology, the soul exclusively as a thinking being ; and, following the Platonic views, he ascribed to it an existence, originally apart from the body, whence it derived as permanent property

Middle Ages.

Return to Platonism.

Descartes as a spiritualist.

all those ideas which go beyond sensory experience. Itself occupying no space, this soul was connected with the body at one point of the brain, in order to receive the influences from the outer world and in its turn to exercise its influence on the body."

Later spiritualism.

The later spiritualism advanced but little beyond this theory of Descartes. Its last great representative was Herbart. He developed in thoroughly logical manner the idea of a simple soul substance, according to Descartes. Herbart was of very great service to the new psychology in a certain way, and we shall say something about his work later (page 284), but his spiritualistic psychology was a total failure. His attempts at deducing the facts of mental life from the idea of a simple soul and its relations to other beings, proved fruitless. His efforts showed more clearly than anything else could do that this pathway was an impossible one for psychology. The idea of a simple soul substance had not been derived from actual observations of mental life but had been arbitrarily and unreasonably asserted ; the facts were to be forced to fit.

Descartes as a materialist.

Descartes contributed to the development of modern materialism in two ways, by his strictly mechanical view of nature in general and by his treatment of animals as automats. Man alone had a mind ; animals were machines. But if the many evidences of thinking, feeling, and willing among animals can be explained physiologically, why cannot the same explanation be used for man ? This was the starting-point for the materialism of the seventeenth century.

Later materialism.

For materialism all facts of thinking, feeling, and doing are products of certain organs in the nervous system. Any observation of the facts of mind is valueless until such facts can be explained by chemical and physical processes. Thinking is a production of the brain. Since this process stops when the circulation of the blood stops and life ceases, therefore thought is nothing but an accompaniment of the materials of which the brain is composed.

"Chemistry of mind."

Down to the present day modern materialism has not gotten beyond this point—mental life is a product of the brain; psychology is merely physiology of the brain. Our feelings, thoughts, and acts of will, however, cannot be observed as all phenomena of nature have been observed. We can hear the word that expresses a thought, we can see the man who formed it, we can dissect the brain that thought it; but the word, the man, the brain—these were not the thought. A feeling of anger is accompanied by an increase of blood in the brain; but no matter how minute our knowledge of the chemical processes between the blood and the brain substance may be, we know that we can never find out the chemistry of anger.

Its impossibility.

But, says materialism, these material processes may not be the thoughts, yet they produce them. Just as the liver produces bile, just as the contraction of muscle causes motion, so are our ideas and emotions produced by blood and brain, by heat and electricity. Yet a very important difference has been overlooked. We can show how the bile is produced by chemical processes in the liver; we can show how the movement is the result of chemical processes in the muscle; but brain processes give us no information of the way thoughts are produced. We can understand how one bodily movement produces another movement, how one emotion or sensation changes to another emotion or sensation; but how a

"Brain produces mind."

Its absurdity.

motion of molecules or a chemical process can produce an emotion is what no system of mechanics can make clear.

Revised materialism.

These vagaries of materialism have called attention to the study of the relations between mind and brain, and we have had "mental physiologies," even from those who are not materialists. The study of what happens in the brain or in any part of the body when we are angry, or when we think of an apple, is, of course, an immensely valuable thing. The absurdity arises when it is asserted that every mental fact is merely an appendix to some brain process; that, for example, we do not feel merry at the thought of a joke, but that certain chemical processes in the brain produced the thought of the joke and at the same time set going other chemical processes that produced the merry feeling. There are many volumes of so-called "psychology" in which each mental process is translated into some imaginary (for we have no facts on the subject) movement of brain molecules, which in some imaginary fashion sets up another imaginary movement, which is translated into a second mental process that really followed the first one according to a simple psychological law.

Fruitlessness of the whole discussion.

But the strife between spiritualism and materialism is almost passed. "It has left no contribution to science, and no one who carefully examines the subject of the strife can wonder at such a result. For what was the central point of the battle of opinions? About nothing else than the questions concerning the soul, its seat, its connection with the body. Materialism here fell into the same fault as spiritualism. Instead of beginning upon the facts that were observed and investigating their relations, it busied itself with metaphysical questions for which answers can be found only—if ever—through a completely unprejudiced—*i. e.*, at the start free from every metaphysical supposition—investigation of the facts of experience."

Mistake of method.

Starting from entirely different points of view, both spiritualism and materialism have landed in utterly fruitless suppositions. The reason therefor lay in the methods which they employed. To suppose that anything could be gained by vague speculation on mental life was folly equaled only by the belief that dissecting brains would lead to a knowledge of mind. Both parties forgot one point—namely, to examine the facts of mind itself.

The new psychology.

It is this forgotten duty that led to the new psychology—a psychology of fact. This psychology of mental life, this science of direct investigation of our thinking, feeling, and doing, is neither spiritualism nor materialism; it has no speculations of either kind to offer. It confines itself strictly to the domain of fact. As long as they can set themselves in harmony with the facts, the Hegelian philosopher and the Feuerbachian materialist have equal rights. When they go beyond the facts, they may settle the question between them; the new psychology is very thankful that it has nothing to do with either.

CHAPTER XXII.

THE NEW PSYCHOLOGY.

Psychology is the science of thinking, feeling, doing.

THE facts we have been considering in this book have been facts of mind, not of the physical world. The beautiful colors we see are—the physicists tell us—only vibrations of ether ; the physical world has no color, the colors exist only when we are present. Physical vibrations of the air are to us tones. Certain mechanical movements are to us pressures. Feelings and will-impulses may betray themselves by movements or otherwise ; in themselves they are mental facts. In short, we may say that all the facts, as we know them, are mental facts. The science of these facts is psychology.

The new method of investigation.

But what is the *new* psychology? The new psychology is entitled to its special adjective because it employs a method new in the history of psychology, although not new in the history of science.

The old psychologist, like Locke, Hamilton, and many of the present day, sits at his desk and writes volumes of vague observation, endless speculation, and flimsy guesswork. The psychologist of the new dispensation must see every statement proven by experiment and measurement before he will commit himself in regard to it. Every alleged fact, every statement, must be brought as evidence—sworn to by the Eternal Truth under penalty of scientific disgrace—before the Court of Reason.

The difference between the old and the new is not one

Difference between the old and the new.

of material; the subject is the same for both, namely, the facts of mind. The difference lies in the carefulness with which the information in regard to these phenomena is obtained. Instead of careless observation and guess-work the utmost care and self-sacrificing labor are expended in the laboratory in order to obtain single facts.

The new psychology declared to be a "bore."

This method of careful, scientific work is unintelligible to the men of the old school. The method of experiment "taxes patience to the utmost" and "could hardly have arisen in a country whose natives could be *bored.*" Just as the schoolmen of long ago were busy in speculating on such questions as, "How many angels could dance on the point of a needle?" so these men write volumes about the Me, the Unity of Consciousness, the Consciousness of Identity, etc. It never occurs to them that the world might ask, "Well, what of it?" Suppose you have proved that 19½ angels can dance on the aforesaid needle. Well, what of it? Suppose you have settled to your satisfaction that consciousness is a unity, is double, or is anything else you please. Well, what of it? When you have written your 500 or 1,000 pages on these subjects, is the world any better off? have you contributed one single fact to the advance of science and humanity? While you were up in the clouds speculating, we were hungering, dying for the lack of information on the most vital questions. Have you no thought, no suggestion as to how we may grow better ourselves and educate our children to a better life?

Psychology at last free.

Who are the men to whom we owe the regeneration? Of course, the psychological awakening is only a part of the great movement by which many of the sciences have successively emerged from the scholasticism of the Middle Ages. Mathematics, physics, chemistry, biology, and others are now free and fruitful sciences ; psychology has just joined the group, leaving education, logic, and æsthetics for some future generation.

English psychology.

Sir William Hamilton is the one to whom we must look back as having vindicated the right to build psychology upon observations and not to deduce it from philosophical prejudices. Since his time English psychology has been nominally empirical (*i. e.*, founded on experience), but actually merely a collection of vague observations as the basis of endless discussion.

German psychology.

In Germany the natural revolt from the dull scholasticism of the psychology of Wolff and the mad speculation of Schelling was led by Herbart.

Herbart.

The philosopher, psychologist, and educator, Herbart, was born in 1776. He became professor of philosophy at Göttingen ; later he succeeded Kant as professor of philosophy in Königsberg, where he died in 1841. He is best known for his works on education ; these being founded on his psychology, have led educational people to adopt the Herbartian psychology with the Herbartian pedagogy. The Herbartian pedagogy, with the improvements of its followers, is, to-

Fig. 203. Johann Friedrich Herbart.

day, perhaps the best system and guide that we have.

To Herbart as a psychologist we also owe a debt. The old faculty-psychology, with its groundless and endless speculation, aroused his ire; he set about producing a new psychology. In the first place, he determined to start from the facts as he observed them in his own mind; this was in itself a great step. You have probably heard of the medieval student who, at the time when the discovery of spots on the sun began to be talked about, called the attention of his old instructor to them. The reply was: "There can be no spots on the sun, for I have read Aristotle's works from beginning to end and he says the sun is incorruptible. Clean your lenses, and if the spots are not in the telescope, they must be in your eye." Positive service of Herbart.

Alas! there are to-day so-called psychologists who write volumes full of what Locke said, what Hamilton's opinion was, what Reid thought, what Hume believed, without ever dreaming of asking what the facts are.

This debt we owe to Herbart is a great one; the other debt we owe him is for a different reason. Mathematics, we all know, is the fostering mother of the sciences. What was more natural than to place poor, disreputable psychology in her care? This is what Herbart attempted. On the basis of his observations he proceeded to build up his mathematics of ideas. His results are very curious; for example, if you have an idea in your mind and another one wishes to get in, there occurs a strife between them and they press against each other with a force proportional to $\frac{a}{\sqrt{a+b}}$. Of course, the whole thing was utterly absurd. Mathematics makes use of symbols for *quantities;* when you speak of a dis- Negative service of Herbart. Algebraic folly.

tance *t*, you mean *just so many* inches or centimeters or miles ; *t* represents a number. But when Herbart speaks of an idea with the intensity *a*, there is no method of giving any quantitative indication of how great this intensity is ; he knows of no measure of intensity, and his use of symbols is absolutely meaningless. No mathematician would ever dream of such folly. The second debt we owe to Herbart is, then, that of a warning example against unfounded speculation. Herbart revolted against metaphysical speculation, but fell into a kind of mathematical speculation that was no less metaphysical.

How to make mathematics applicable.

But if all that was lacking was merely the quantitative expression for psychological facts, why not get to work and measure them, just as in astronomy and physics? But how? How can we measure the intensity of a pain, or the time of thought, or the extent of touch? The matter seems really incomprehensible.

Herbart and Fechner.

One of the surest ways of being put in the wrong is to say that something can never be done. Comte, the philosopher, once said that it would be forever impossible to tell the composition of the stars ; forty-three years later the use of the spectroscope enabled astronomers to analyze each one. Herbart declared that "psychology must not experiment with man ; and instruments thereto do not exist" ; in another place he asserts that "psychological quantities are not presented in such a way that they can be measured ; they allow only an incomplete estimate." Nineteen years later Fechner published his great work on psychophysics, in which he showed how to experiment on mental processes and measure psychological facts.

New psychology is a development.

Other influences had been tending toward the development of psychology, and, although Fechner was the first really to start the new psychology, he is only the logical outcome of the progress of thought in other lines.

Influence of physics and physiology.

Both the physicist and the physiologist frequently come to problems where mental life is involved. Physicists still amuse themselves by the so-called optical illusions and the beautiful phenomena of contrast, although there is not a particle of physics in any way connected with the subject. Physiologists have always been forced to consider questions of sensation, emotion, and volition, in order to draw conclusions in regard to bodily processes. Many names might be mentioned in this connection, but one is of special importance, that of Ernst Heinrich Weber. This distinguished physiologist and physicist wrote a semi-psychological treatise on "Sensations of Touch and the Internal Feelings," which not only induced later physiologists to continue the work, but was also the direct stimulus for Fechner. This influence we may call the physiological one; it has done its main psychological service in outlining the sensations in a qualitative manner. Fechner may be considered as the builder of psychology representing the

Weber.

Fig. 204. Gustav Theodor Fechner.

final passage from the qualitative to the quantitative.

Fechner.

Fechner (1801–1887) was the founder of experimental psychology. While professor of physics at the University of Leipzig he invented and worked out the methods which we have used in finding the threshold (page 103, etc.). His greatest works were, "Elemente der Psychophysik" and "Revision der Hauptpunkte der Psychophysik." So much of Fechner is embodied in all our psychological work that it is useless to attempt more than to indicate his main services. I will sum them up as: (1) the invention of new methods of measuring the intensity of sensation; (2) the introduction of new methods of calculating results; (3) the development of laws concerning the relation of intensities of sensations; (4) the foundation of experimental æsthetics; and (5) numerous smaller investigations and observations.

Fig. 205. Hermann von Helmholtz.

Helmholtz.

A greater than Fechner was to come. Mathematician, physicist, physiologist, psychologist, and technologist, Hermann von Helmholtz has given to the psychology of sight and hearing the best his sciences had to give. We cannot claim him as a psychologist, his genius was too great for a science still so limited. Nevertheless there are few to whom psychology owes more.

Influence of astronomy. We must turn back to the last century for a second current of thought that was to develop psychology. This time it was an astronomer puzzled by mistakes of his own method. In a preceding chapter (page 40) the story has been told. The time measurements of mental phenomena were afterwards taken up and developed by Wundt, in whose laboratory they are still continually pushed further.

Wundt. Wilhelm Wundt, born at Neckerau in Baden in 1832, was a student of medicine at Tübingen, Heidelberg, and Berlin. His academical career began with a place as instructor in physiology at Heidelberg, where in 1863 he published his "Lectures on Human and Animal Psychology" (2d edition 1892). In 1864 he was made assistant professor of physiology. In 1866 he published "The Physical Axioms and their Relations to the Principles of Causality." In 1874 he published the "Outlines of Physiological Psychology" (4th edition 1893). In the same year he was called to Zürich as professor of philosophy;

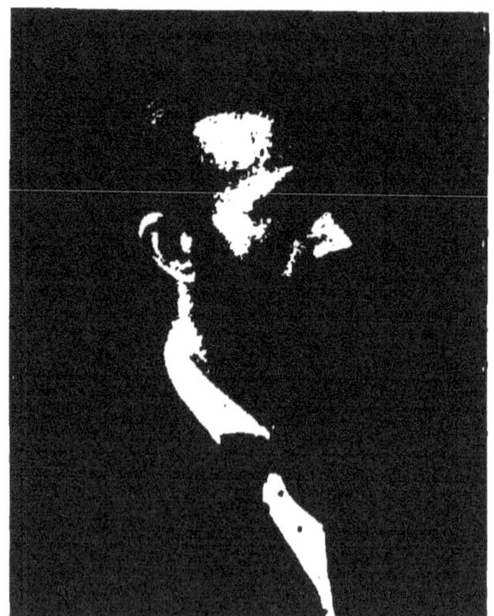

Fig. 206. Wilhelm Wundt.

in 1875 to Leipzig. His later works have covered most sections of philosophy: "Logic" (1880, 1883, 2d edition 1893), "Essays" (1885), "Ethics" (1886, 2d edition 1894), "System of Philosophy" (1891). The productions of the Institute for Experimental Psychology at Leipzig are published in his "Philosophical Studies."

Institute at Leipzig.

The institute at Leipzig has taken up not only the time measurements and the work begun by Fechner, but also nearly every portion of psychology accessible to experiment. I think it can be said that there are only two important regions of psychology which have not received contributions from Leipzig, namely, power and work in voluntary action (investigations from France and Italy) and the applications of psychological principles to education (a peculiarly American department). When we view the ten solid volumes of investigation in the "Philosophical Studies" and remember that the men who produced them were simply carrying out Wundt's thoughts, we must admit the justness of the recognition which the world pays to Wundt as the greatest genius in psychology since the time of Aristotle.

Influence of anatomy.

There is another influence on the development of psychology, which we might call the anatomical. Arising from crude materialism, it has sought to parcel out mental life among different portions of the brain. It has produced such monstrosities as "mental physiology," "physiology of mind," etc. It speaks of the different mental faculties as seated in various portions of the surface of the brain.

Anatomical speculation in psychology.

The evil this movement has done is very great. It has led to a habit of wild speculation concerning "memory cells," "association fibers," etc. Concerning what goes on in the brain in company with mental processes we know nothing more than general outlines. Yet writers of psychologies—especially in America—generally disdain to speak of the relations of mental processes to each other. They first turn the sensations into "molecular movements" (this is a favorite phrase because its meaning is so deliciously indefinite and its alliteration so sonorously professorial); then they set complicated processes running along nerve fibers to other

Fig. 207. Lecture-room in the Yale Laboratory.

cells; here there are more "molecular movements," which are retranslated into mental phenomena. A mental process would be very likely to suffer from so many translations, especially when the intermediate processes are, as the anatomist will tell you, utterly unknown languages.

As would be expected, it is not the brain anatomist who commits this fault. He is after facts concerning the structure of the brain. Experimental psychology can be of great service to him, especially in the pathology of mind, where brain disease is accompanied by mental disturbance or mental troubles disastrously affect the body. These men are ardent and valuable friends of our science.

Three stages of speculation.

Indeed, the subject of psychology has passed through three stages of metaphysical speculation: the doubly distilled metaphysical speculation of the pre-Herbartian

Fig. 208. Apparatus-room and Switchboard-hall in the Yale Laboratory.

era, the mathematico-metaphysical speculation of Herbart and his followers, and the anatomico-metaphysical speculation of Carpenter, Maudsley, and the rest.

Psychology and philosophy.

And what about philosophy, the science of sciences? Alas! philosophy is still in the Middle Ages. One by one the other sciences have freed themselves; the lingering clutch of philosophy on psychology is a last hope of respectability. Metaphysics of the worst sort still goes begging for recognition under such terms as "rational psychology" (as though the psychology of fact were irrational!), "theoretical psychology," "speculative psychology," etc. Philosophy in modern times has contributed nothing but stumbling-blocks in aid of psychology.

Psychology has no connection with philosophy.

The trouble lies in the fact that the new philosophy has only lately appeared. The movement by which a science of philosophy is to be based on the special sciences, such as physics, psychology, and the rest, is still so new as to be very little known outside of Germany. This new philosophy has no more and no less connection with psychology than with physics, mathematics, and astronomy. Every scientist, every man, must be more or less of a philosopher in the new sense, and philosophy based on the special sciences cannot but be a help in every way. Such a philosophy would no more think of claiming a right to meddle in psychology than it would to regulate the manufacture of lathes in a machine shop.

View of the present.

Having arrived at the present day, we naturally ask, What is going on now? In Germany the number of psychological laboratories is not large and the amount of capital invested is small. Yet, it must be confessed, the best work and almost all the good work in psychology comes from the German psychological, physical, and physiological laboratories. The causes combining to this result are many; the main one is hard, honest, accurate work.

Germany.

America.

In America the first laboratory was founded at Johns Hopkins University in 1883 by G. Stanley Hall. The work done was excellent and full of promise, but the laboratory was allowed to pass out of existence upon the departure of Professor Hall. At the present moment, there are about twenty American institutions in which attempts are made at laboratory instruction.

France has lately established a laboratory. Russia

Fig. 209. Workshop in the Yale Laboratory.

has a laboratory at Moscow. Up to the date of my latest information no laboratories exist in Austria, Italy, Spain, British Empire (except Canada), or elsewhere.

Yale. The Yale laboratory was started in 1892. It differs from the German laboratories in having an organized system of courses, whereby the college student receives a thorough training. It differs from most American laboratories in its extensive provisions for accurate, scientific researches by trained investigators of special subjects.

Laboratories are the outward signs of internal forces at work in developing psychology. Forces of development.

The first of these forces is the conviction in the mind of every man that mere observation and speculation will not serve to build a locomotive, paint a picture, run a gas factory, or teach psychology. Long, long years of special training and laborious experimenting must first be spent in the workshop, the studio, the chemical laboratory, or the psychological laboratory. To do any of these things a man must be a specialist. As long as psychology was an arm-chair science, anybody could teach it; to-day no one but a carefully trained man can do so. Power of the specialist.

A second great force is the recognition that all rational and effective education is based on psychology —not the vague and verbose "psychology" of ten, twenty, or forty years ago, but the accurate, up-to-date, practical psychology of to-day. Basis of education.

INDEX.

www.ingramcontent.com/pod-product-compliance
Lightning Source LLC
Chambersburg PA
CBHW021506210326
41599CB00012B/1146

* 9 7 8 3 7 4 3 4 0 2 7 4 4 *